Multimedia Information Retrieval

Matthias Hemmje · Stefan Wagenpfeil

Multimedia Information Retrieval

Extracting Information from Multimedia Sources

 Springer

Matthias Hemmje
FB Informatik FernUniversität Hagen
Hagen, Germany

Stefan Wagenpfeil
PFH Private University of Applied Sciences
Göttingen, Germany

ISBN 978-3-662-73309-7   ISBN 978-3-662-73310-3   (eBook)
https://doi.org/10.1007/978-3-662-73310-3

Foreword

Developments in the fields of multimedia and artificial intelligence are closely interconnected in many respects. Therefore, it seems appropriate to have the foreword for a multimedia information retrieval book written by an artificial intelligence specializing in text generation:

"Multimedia Information Retrieval (MMIR) is a field of information technology that deals with the organization, search, and multimodal data such as text, images, videos, and audio files. The purpose of MMIR is to facilitate information retrieval and analysis and to increase efficiency when working with large volumes of data.

Over the years, various technologies and algorithms have been developed in the field of MMIR to address this challenge. One of the most important methods is feature extraction, which provides a representative description of the content of a multimedia file. These features are then stored in a database and can serve as the basis for subsequent search and organization of the data.

Another important method in the field of MMIR is semantic search. In this approach, a semantic model is created that describes the relationships between the various elements of the content in a multimedia file. These models can be used to facilitate the search for specific content within a database and to identify connections between different elements of the content.

In practice, MMIR is applied in various fields, such as medicine, education, or the entertainment industry. In medicine, for example, MMIR technologies can be used in the diagnosis and treatment of diseases by facilitating the analysis of medical images and audio files. In education, they can be used to create e-learning platforms or to personalize learning content.

In the entertainment industry, MMIR technologies are increasingly being used to automatically recommend music and videos based on user preferences. The recognition of faces in movies or the analysis of social media posts are also among the application areas of MMIR.

Overall, the goal of MMIR is to increase efficiency in the search and organization of multimodal data and to enable users to gain a better understanding of and greater control over such data. With the growing prevalence of multimedia content, MMIR will continue to play an important role in the future, further facilitating information retrieval and analysis."

May 2025   Chat-GPT
                                                                          (im Auftrag der Autoren)

Preface

Multimedia Information Retrieval

"Multimedia is everywhere" [1]—there is no better way to summarize today's IT landscape. Not only is an almost incomprehensibly large amount of multimedia content produced, edited, shared, and consumed every day, but nearly every application and interaction between humans and machines relies on multimedia forms of presentation. Fig. 1 impressively illustrates how this digital world took shape on the Internet in 2023. On the Internet alone—without considering internal company information, private archives and activities, or offline-generated media—2.4 million emails are sent every second, 54,319 Google searches are processed, over 120,000 videos are watched on YouTube, and nearly half a million WhatsApp messages are sent. Every second, 721 people share their food on Instagram and 28,935 people "like" it. All of this happens every single second!

The year 2023 marks a watershed moment in many respects. On the one hand, the global coronavirus pandemic had a massive impact on the use of various multimedia technologies; on the other hand, the release of the first artificial intelligence applications ushered in a new era. Since then, not only humans but also automated systems have been able to generate multimedia content in deceptively realistic quality. This has led to an exponentially greater growth in multimedia content, but it also means that statistics such as those in Fig. 1 are now difficult to restrict to content produced by humans.

We all know that the number of multimedia contents and multimedia users continues to rise. And we have also noticed that the "Level Of Detail," that is, the level of detail in multimedia, is also increasing. Videos are now produced in 8K resolution, smartphone cameras have over 100 megapixels [3], and thanks to ever faster internet connections, all of this can be produced, distributed, and consumed at very high speeds. However, the challenges this poses for backend systems, where multimedia is managed, organized, distributed, or converted, are enormous. But even in traditional standalone applications, multimedia has become indispensable. Every computer game, every streaming service, social media, every interaction—in fact, every single button that an application displays to you somewhere—is a multimedia interaction. The animation when pressing the button is

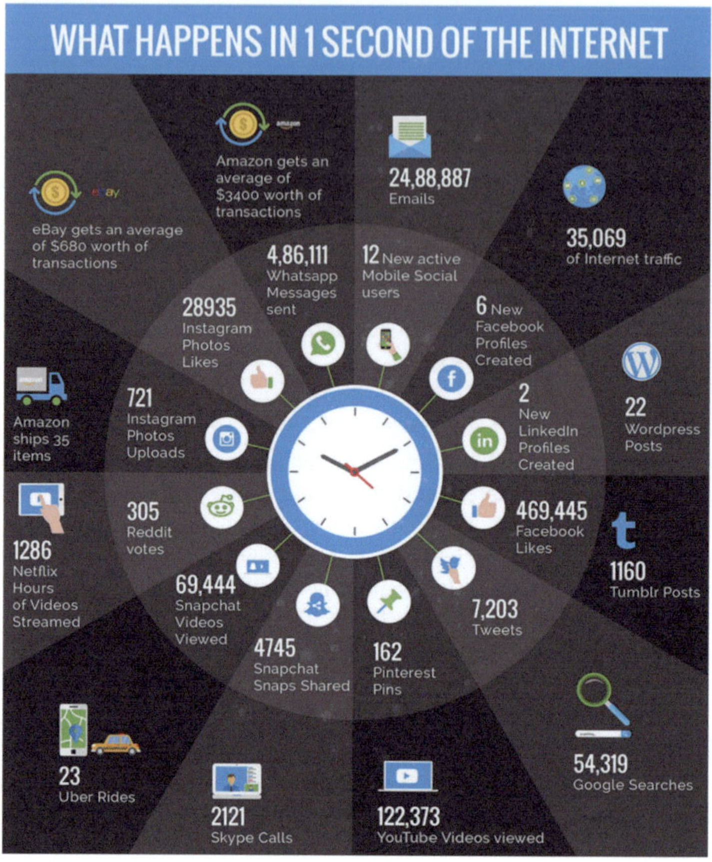

Fig. 1 One second on the Internet (adapted from [2])

multimedia, but even deciding what to display next can be considered a classic multimedia information retrieval process. Thus, it is not even necessary at first to bring all the new technologies from the fields of Virtual Reality/Augmented Reality/Metaverse into play to highlight the relevance of multimedia information retrieval. However, these areas do, of course, very strikingly demonstrate where, how, and why multimedia is important for every type of application.

In this textbook, the terms, concepts, and methodologies in the field of "Multimedia Information Retrieval" are explained, formally introduced, and illustrated with practical examples.

In the following sections, you will find various visual elements. These are intended to help you, for example, to review the most important points at the end of a subchapter, maintain an overview, and reflect on what you have learned.

> The key points of a subchapter are highlighted in a color-coded block.

At the end of each subsection, you will find a series of self-assessment exercises that allow you to check the knowledge you have already acquired. The corresponding solutions are collected at the end of the book.

However, you will also find a large number of examples and source codes in this book. These are numbered, also specially formatted, and referenced in the corresponding text (see Listing 1).

```java
1    public class HelloWorld {
2            public static void main(String[] args) {
3                    System.out.println("Hello_World.");
4            }
5    }
```

Listing 1 Example of source code formatting in the book.

Technical commands, class names, packages, or designations are printed in proportional font, important terms, names, or concepts *are formatted in italics*. References for the corresponding passages can be found at the end of the book. There are a total of six chapters:

- **Chapter 1—Information and Knowledge:** This section introduces fundamental concepts necessary for performing information retrieval. In particular, various formalizations of the concepts of information and knowledge are presented, which will play an important role throughout the rest of the book.
- **Chapter 2—Features and Feature Representation:** This section examines the various multimedia types and their features, which are then processed in information retrieval systems.
- **Chapter 3—Information Retrieval (IR):** This part introduces the components, models, properties, and metrics for classical information retrieval.
- **Chapter 4—Models and Concepts of IR:** An ever-growing volume of data must be managed, organized, distributed, converted, and archived. Above all, access to information must be fast and efficient. This subsection presents the current methods and standards for achieving this.
- **Chapter 5—Evaluation and Relevance Metrics:** The assessment of the methods presented so far requires reliable evaluation techniques and appropriate relevance measures to determine the quality of IR.

- **Chapter 6—Applications and Outlook:** The book concludes by placing a range of current multimedia applications, generative AI, and virtual reality in the context of multimedia information retrieval.

For reasons of readability, the generic masculine is predominantly used in this book. This always implies both forms and therefore includes the feminine form as well.

Matthias Hemmje
Stefan Wagenpfeil

Acknowledgments

This book is based on content from the courses "Multimedia Information Systems" and "Information Retrieval" by Prof. Dr.-Ing. Hemmje at the Department of "Multimedia and Internet Applications" at FernUniversität Hagen, as well as the lecture "Information and Knowledge Management" by Prof. Dr.-Ing. Wagenpfeil at the Department of Business Informatics/Software Engineering at PFH Göttingen, and meaningfully supplements these.

Several esteemed colleagues, staff members, and students contributed to the development of the courses and lectures on which this book is based, and we owe them our gratitude: for the courses at the FernUniversität in Hagen, these are Arnd Steinmetz, Birgit Ianiello, Holger Brocks, Gunter Sterr, Carsten Bornstein, Sadric Vid, Frank Lehmann, Peter Spindler, Douglas von Roy, Arne Schulz, Thomas Klinkert, Stefan Einer, and Felix Engel.

Contents

List of Figures

List of Tables

Information and Knowledge

1

This first chapter deals with the different definitions of the terms "information," "knowledge," and "data." You will find that there are a variety of approaches here, each focusing on a particular aspect of information flow, knowledge exchange, or data transmission. First, the more general information models based on Nonaka and Takeuchi [2] are introduced (see Subsection 1.1), which focus heavily on the concepts of externalized and internalized knowledge. Subsequently, Subsection 1.2 presents the well-known model by Kuhlen 1.2, which contributes a number of important definitions. The description of a user's information need is introduced using Belkin's ASK model 1.3 and further concretized by several additional concepts regarding information acquisition. In Sect. 1.5, a comprehensive interaction model is derived based on the theories presented, which will be revisited throughout the rest of the book. Finally, the subsection on information models 1.4 demonstrates how interactions between users or between a user and the system can be formalized and modeled.

1.1 The SECI Model

A good starting point for delving into the world of information and knowledge concepts is the work of Nonaka and Takeuchi [2], which provides an important foundation for many theories and models. In their publications, the two authors coined the SECI model, which consists of four interlocking phases (see Fig. 1.1): Socialization, Externalization, Combination (English "Combination"), and Internalization. The following sections explain each phase in more detail. In the further course of the book, a formal designation is also used to assign individual processes to the SECI model, representing the respective phase. For this purpose, the Greek letter *Sigma* (σ) is used as an identifier, and each phase is denoted by its initial letter: σ_S for Socialization, σ_E for Externalization, σ_C for

© The Author(s), under exclusive license to Springer-Verlag GmbH, DE, part of Springer Nature 2026

M. Hemmje and S. Wagenpfeil, *Multimedia Information Retrieval*, https://doi.org/10.1007/978-3-662-73310-3_1

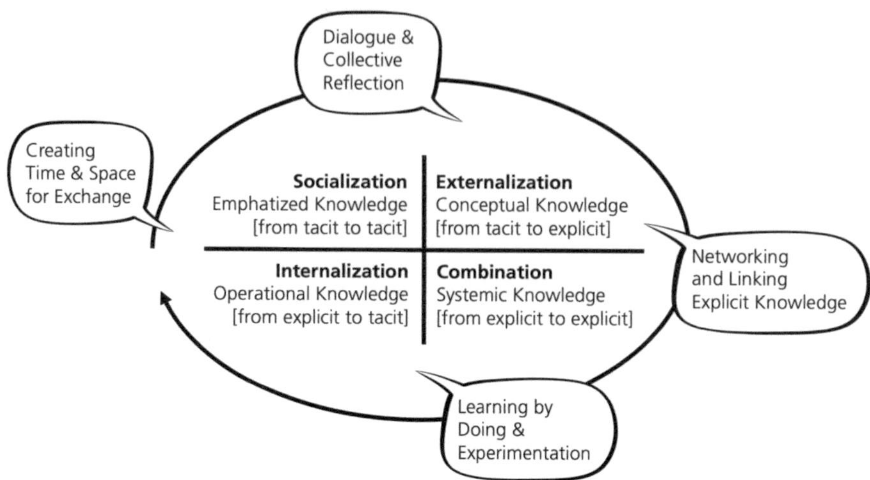

Fig. 1.1 The SECI information model. (Own illustration based on [2])

Combination, σ_I for Internalization. A complete SECI cycle can thus be described as $\sigma = \{\sigma_S, \sigma_E, \sigma_C, \sigma_I\}$.

This first phase, "Socialization" σ_S, is probably the most original phase of knowledge processing. Here, knowledge exists in an internal form, i.e., only in the "mind" of the user. There is no written record, no documentation, no textual or visual description of the knowledge; it is passed on solely through demonstration. This phase could be described as "training on the job," for example, when an experienced user demonstrates to an inexperienced user by "showing and doing" how to achieve the desired results, thereby imparting their knowledge. This process is, of course, very time-consuming and not very sustainable, as it always requires an explicit social relationship between the original owner of the knowledge and the recipient. Therefore, appropriate time must be allocated for such a knowledge exchange.

The σ_E "Externalization" phase now introduces documentation or the written recording of knowledge. The group of people possessing knowledge begin to prepare their knowledge in such a way that it becomes conceptualized. That is, it is written down, explained, documented, illustrated, etc. At this point, of course, all modern forms of knowledge documentation can be incorporated—the crucial point is that the knowledge is no longer only in the "mind" of the user but has been externalized into a form that can be further utilized. At this stage, all mechanisms typically used to produce suitable documentation also come into play: requirements analysis, target group analysis, discussions, reflection on key points, definition of the documentation format, translations, and much more.

Externalized knowledge is generally quite suitable for being combined at a systemic level in the "Combination" phase σ_C. This area includes expert systems, databases, business intelligence systems, machine learning, and, of course, all information retrieval (IR) systems (note: a precise definition of IR will follow in later subsections; for now, it suffices to consider IR as the process of finding and accessing information). The main task in this phase is to link knowledge, possibly derive new knowledge, and also externalize it. It is important to note that this phase is usually purely technical, as the user typically only becomes involved again in the next phase.

The "Internalization" phase σ_I ultimately describes the (re-)absorption of knowledge. Externalized knowledge, i.e., knowledge that exists in some form of documentation, media, or systems, should, of course, also be made available again for solving problems or answering queries. The moment a user receives and perceives a response from an information system, this knowledge is internalized again, i.e., it moves from the information system, document, or medium back into the "mind" of the user. At this point, the exploration of new knowledge is also supported, for example, when the user researches in archives or examines the results of experiments and possibly draws conclusions from them.

A crucial foundation for further engagement with knowledge and information theory is to clearly state in which phase a particular exchange of knowledge takes place. Knowledge exchanged, for example, between humans and machines—referred to as "Human-Computer Interaction (HCI)" [3]—typically occurs in the phases σ_S-σ_E or σ_C-σ_I, whereas a conversation between two people can usually be described solely by the S phase. Another important area is "Computer-Supported Cooperative Work scenarios (CSCW)" [4], in which the work—and thus indirectly the knowledge—of users is supported by the use of information systems. These scenarios also take place in the phases σ_S-σ_E and σ_C-σ_I. Pure human-human interaction (HHI) takes place in the σ_S phase. All of this can also be represented as a mathematical function. In the following syntax, an arrow denotes the exchange of knowledge and a phase transition in the SECI model. Thus, the notation $\sigma_S \rightarrow \sigma_E$ can be understood as the transition from the socialization phase (we talk to each other) to externalization (we write it down). Of course, writing it down can also be done digitally using a computer. In addition, symbols are used to link functions, such as the "or" symbol $\vee$.

The three scenarios described in this subsection—HCI, CSCW, and HHI—can therefore be formally described based on the SECI model as follows:

$$\text{HCI:} \sigma_S \rightarrow \sigma_E \vee \sigma_C \rightarrow \sigma_I \tag{1.1}$$

$$\text{CSCW:} \sigma_S \rightarrow \sigma_E \vee \sigma_C \rightarrow \sigma_I \tag{1.2}$$

$$\text{HHI:} \sigma_S \rightarrow \sigma_S \tag{1.3}$$

You may have noticed so far that the discussion here has mostly been about "knowledge" and rarely about "information" or "data." To clearly distinguish these two terms, Kuhlen [5] has formulated a number of important definitions, which will bring more clarity to the terminology in the next subsection.

> The SECI information model consists of the phases Socialization, Externalization, Combination, and Internalization, thus forming a fundamental model for information exchange.

1.2   Information and Knowledge

Kuhlen [5] coined a very famous phrase: "Information is knowledge in action." This statement summarizes his collected works very well and at the same time introduces a new perspective on the concepts of information and knowledge: action. This phrase provides an excellent basis for distinguishing between "information" and "knowledge." Knowledge is essentially something general that simply exists. Only when a specific action takes place can knowledge become information. The underlying idea is that each of us—and technically speaking, every information system as well—accumulates a vast amount of knowledge that is stored in some form. Most of the time, however, this knowledge is not relevant. Only when a specific action occurs can part of this knowledge become relevant and thus become the required information.

A simple example: We all know our birthday. This is stored as knowledge in our brains. Normally, however, knowledge of one's birthday is rarely used. But if you have to fill out a form or want to apply for a new ID card and are asked for your birthday, this knowledge suddenly becomes highly relevant because it is needed—in the context of the action you are performing. And thus, knowledge becomes information. If you share this information with someone else or even write it down, you can immediately see the analogies to the SECI model (see Subsection 1.1): if you write down your date of birth, you are externalizing your knowledge. If you simply tell it to the person asking, we are in the socialization phase. If your date of birth is stored or matched with information systems, that would be combination, and the person who, for example, found your date of birth in an information system (without asking you) would then internalize this knowledge.

It is also interesting that in human-human interaction, action is required on both sides for a knowledge exchange to become an information exchange. You could stand in a pedestrian zone and tell random passersby your date of birth (an action from your perspective), but the knowledge exchange only takes place when a passerby also wants to know and learn when your birthday is (and that would be the action from the passerby's perspective).

Through this distinction by Kuhlen [5], it also becomes clear that when modeling systems, a difference must be made between pure knowledge and actual information. If we imagine knowledge as a large database, vast amounts of records can be accumulated and stored there. According to Kuhlen, searching a database would then be the necessary action that extracts from these vast amounts of data those that represent a certain added value as information for the user. Especially with exponentially increasing data volumes and ever greater levels of detail, considering the action and its associated relevance is essential. Knowledge and data can be equated in a technical context. In a more human context, the term "knowledge" would be preferred, while in a more technical context, the term "data" would be used; for example, databases fall into the category of knowledge-based systems.

Kuhlen has thus provided a good definition of the difference between knowledge and information, but the action that turns knowledge into information is still relatively unclear. A very nice model that fills this gap comes from Belkin [6] and will be introduced in the next subsection.

> The distinction between knowledge (factual data) and information (currently relevant, i.e., needed knowledge) is an important prerequisite for modeling information processes.

1.3 Information Need

The action introduced by Kuhlen, in the course of which knowledge is needed and thus becomes "relevant," often reflects an aspect that can also be called information need. This concretizes a rather general action in relation to the information required for the user to successfully perform this action. By modeling the information need as a separate term, the quality of the information can later be evaluated: Was it correct, was it complete, did it help the user successfully carry out their action? Belkin [6] approaches this information need by describing required knowledge, i.e., missing information, as an anomalous state of knowledge (ASK), and it is the task of information retrieval systems or informational interactions to resolve this anomalous state. In this way, Belkin directly models the missing information and thus exactly what is required from an IR system as a result. If the system provides an incorrect or incomplete result, the ASK cannot be resolved, and the information need remains wholly or partially unmet. It is also interesting that Belkin thus models that a user must first recognize that an ASK exists, from which their information need can then be derived. And—to bring Kuhlen into this discussion—from which an informational action arises, aimed at obtaining information that will then eliminate the ASK.

1.3.1 Formulating the Information Need

Of course, in the field of multimedia, there are also different user groups with completely different levels of knowledge, as well as various applications that, based on this knowledge, provide options for formulating the information need. Even clicking on a link can be considered an indirect formulation of an information need by users, since they obviously have the desire to access the information on the page linked. Users without knowledge of how to formulate an information need are typically "guided" by applications through menu structures or input fields. A good example of this is social media applications, which are essentially multimedia information systems (MMIS) and multimedia information retrieval (MMIR) systems, but which elicit the user's information need through simple functions such as "Follow" or "Like." MMIR functions are cleverly hidden behind an intuitive menu structure, so the applications can be used without prior knowledge (see Fig. 1.2). This type of application requires no knowledge of MMIR from the users. On the other hand, users can only formulate their information need in the ways provided by the application's developers.

Search fields within an application, where queries can be formulated via keyword search, are also widespread, easy for users to operate, and often provide a good way to specify the information need more precisely through additional filtering options (see Fig. 1.3). For the user, it is not clear how the MMIR system formulates a query from their keywords, nor is it often clear why certain items appear in the results list (e.g., the building in the second row of Fig. 1.3), yet users are obviously quite satisfied with this type of use and the accuracy of the results—as the success of search engines like Google also demonstrates.

Applications that implement the query-by-example pattern are also very easy for users to operate. Here, the user can formulate their information need by providing an example. Figure 1.4 shows Google Reverse Image Search, where any photo (left) can be used as the starting point for search processes. The results found are then displayed on the right. This pattern is often combined with options for query refinement, which in the example mentioned is accessible via the "Are you satisfied with this result" option in the lower right area.

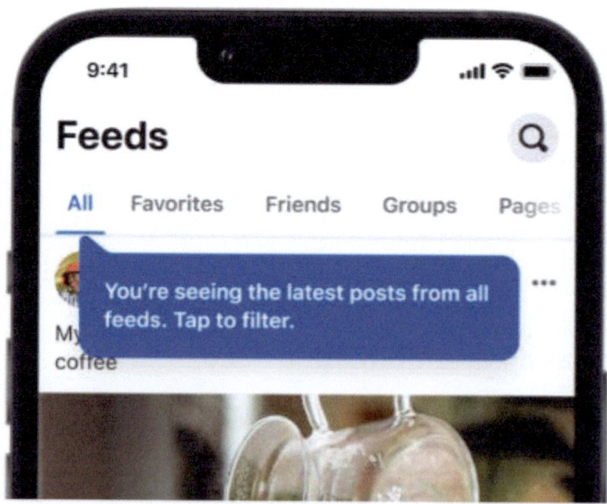

Fig. 1.2 Indirect formulation of information need in the Facebook app [7]

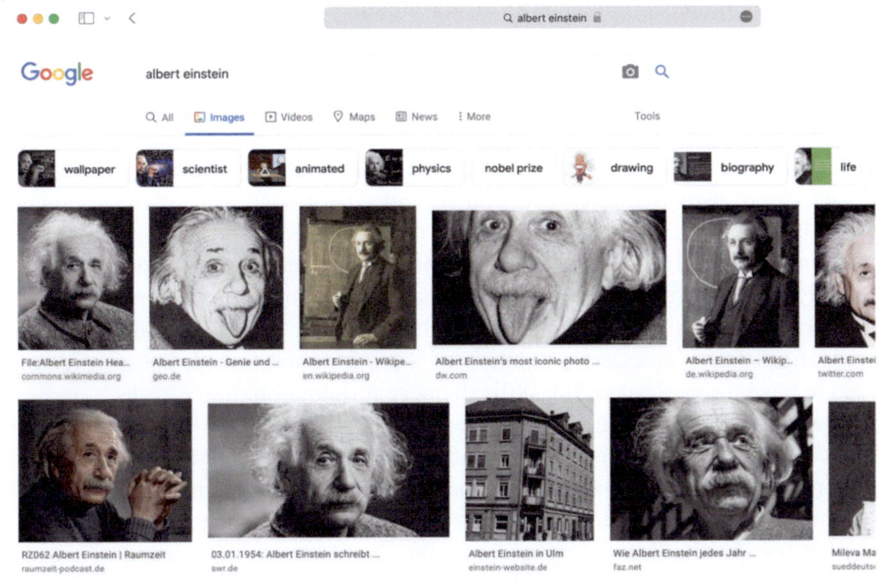

Fig. 1.3   Google keyword search for images with "Albert Einstein" [8]

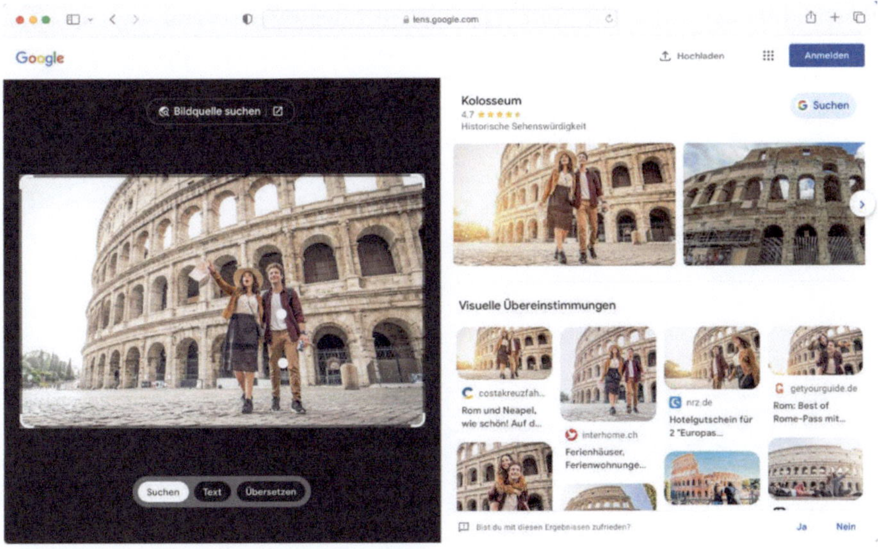

Fig. 1.4   Google image search by providing an example [8]

The next, more complex level of formulating information needs is reached when users are given options to specify their query more precisely in addition to a simple, keyword-based search. However, this often requires users to have more knowledge about formulating queries and, in many cases, to be more familiar with the application's information model. In multimedia archive solutions, syntactic extensions are often provided in the search fields, allowing users to formulate Boolean expressions, constraints, and conditions, such as the following: `+((+Obama +Barak) -Michelle) +(loc:New York)))`. This query, for example, would search for results annotated with the words Obama and Barak, but only if the word Michelle is not included in the annotations. In addition, the extra condition is specified that the location (`= loc`) must be New York. This allows for a fairly precise formulation of the information need, but also requires users to have the corresponding skills in query formulation.

The most complex, albeit most powerful, MMIR systems provide users with full access to the data base in the form of structured query languages. Queries can (and must) then be formulated by the user in a special query syntax and sent to the system. Using these applications requires not only knowledge of query formulation but also of the internal structure of the data base, as knowledge of the data model and schema is often needed to formulate valid queries at all. Common queries include SQL [10], GraphQL [11], SPARQL [12] (see also Fig. 1.5)

More complex ways of formulating the information need require users to have more knowledge. Users must first acquire this knowledge in order to use such applications. There are often different types of users who also operate applications in different ways. Therefore, applications often offer several ways in which information needs can be formulated. In addition to the example of the "Like"

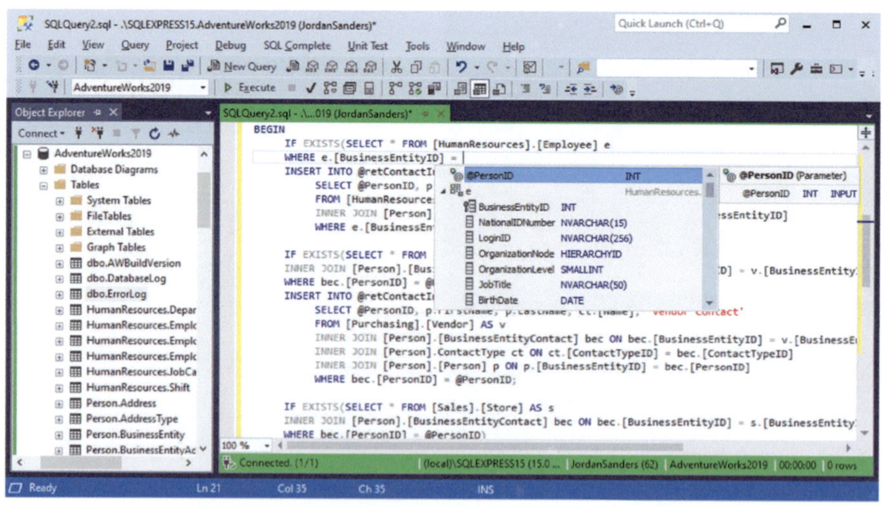

Fig. 1.5   SQL Server Management Studio [13]

button, which can be used immediately and correctly by any type of user without additional knowledge, Facebook, for example, also provides a complete software interface (Application Programming Interface, API) that allows trained users (in this case, mostly developers) to fully formulate their information needs. For this user group, the user interface would no longer look like Fig. 1.2; instead, queries can be formulated in technical formats such as JavaScript Object Notation (JSON) [14] and submitted to the Facebook MMIR component via technical services such as SOAP or REST services (Simple Object Access Protocol, Representational State) [15, 16] (see Fig. 1.6).

Opportunities to formulate the information need using natural language querying are also increasing. In addition to simple queries that can be made to digital assistant devices (e.g., "Hey Siri") and usually produce simple results, in recent years the use of generative AI systems has increasingly been used to formulate the information need. Chat-GPT [18], for example, offers users the opportunity to interact with an MMIR system in natural language and in this way formulate the information need and receive the results, i.e., to read or even hear them via speech synthesis. Regardless of the way the information need is formulated, there are a number of general models for accessing information, which will be explained in more detail in the following subsection.

> The information need is a central concept in modeling information flows. It serves as the starting point for many IR processes.

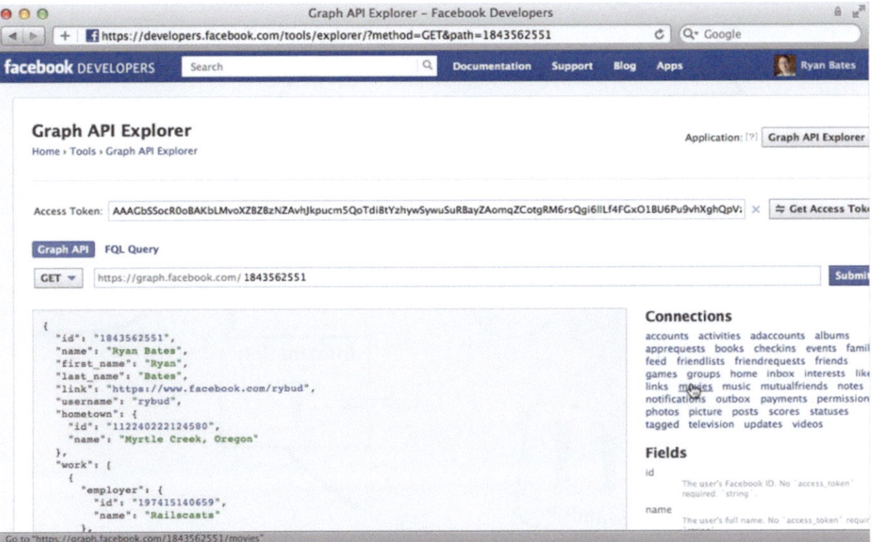

Fig. 1.6 Facebook Graph API [7]

1.4 Information Models

The previous considerations already lead to an initial information model from the user's perspective. This is visualized in Fig. 1.7 and serves as the basis for defining some terms and activities in the following subsections.

First, Fig. 1.7 shows that on the user side we talk about knowledge, whereas on the system side (right) the term data is used. Most of this book deals with the system side, but in this first step we want to clearly define the users' perspective on such a system. The architecture and structure of such information systems can be found in Section 3.2.

A central aspect in the formalization of information and knowledge exchange is the formulation of the information need I_{Need}. Equally central is the representation of the result I_{Result}, which is determined in response to an information request I_{Query}. Information requests are formulated in special query languages $q \in Q$, which are interpreted by the respective information systems. The information need, in turn, can be derived from the user's ASK, which is determined based on the set of existing knowledge W and considering an action act from the set of all possible actions A, or a task or decision. The result of a query r is based on the data D available in the external information system. These facts can then be represented a bit more formally and mathematically as follows: $\forall ask \in ASK \notin W, act \in A, w \in W, q \in Q, r \in D$:

$$I_{\text{Need}} = act \to ask \tag{1.4}$$

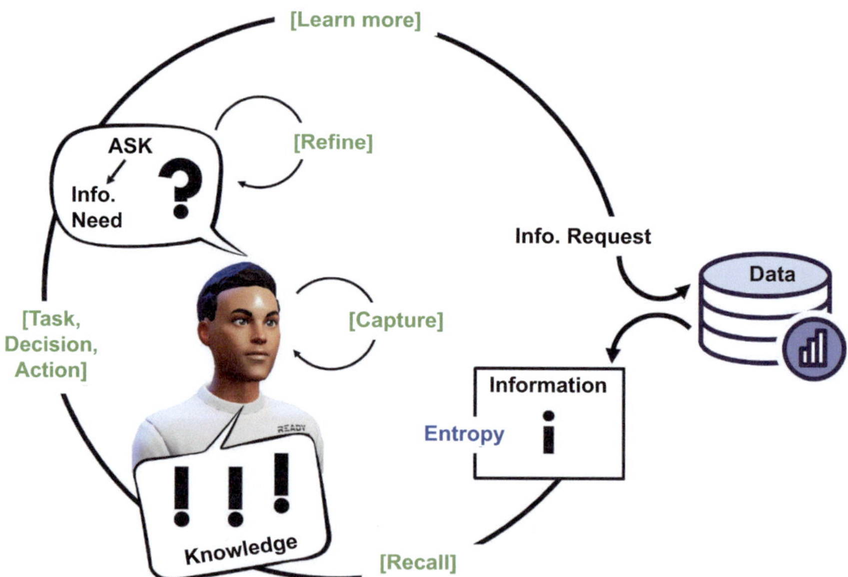

Fig. 1.7 Information and knowledge from the user's perspective

$$I_{\text{Request}} = I_{\text{Need}} \rightarrow I_{\text{Query}} \tag{1.5}$$

$$I_{\text{Query}} = ask \rightarrow q \tag{1.6}$$

$$I_{\text{Result}} = q \rightarrow r \tag{1.7}$$

In general, the process can of course also be classified using the SECI model:

$$I_{\text{Need}} = \sigma_S \rightarrow \sigma_S \tag{1.8}$$

$$I_{\text{Request}} = \sigma_S \rightarrow \sigma_E \tag{1.9}$$

$$I_{\text{Query}} = \sigma_E \rightarrow \sigma_C \tag{1.10}$$

$$I_{\text{Result}} = \sigma_C \rightarrow \sigma_I \vee \sigma_E \rightarrow \sigma_I \tag{1.11}$$

The result r returned by the information system must now be further examined. The result initially comes from a "foreign" data or knowledge base and was determined by the information system as best as possible. The information system has thus automatically calculated that, with this set of information, it can successfully satisfy the user's query and thus their information need. However, this is not always the case in practice. For example, users may not have formulated their queries correctly and thus receive results that do not match their ASK. It may also be that the results are simply insufficient to satisfy the information need due to the data base in the information system. Therefore, the result must first be evaluated from the user's perspective. A measure of the relevance of the retrieved information for a specific user is the entropy H. It can be formalized mathematically and thus calculated as follows: as the remaining ASK considering the result set.

$$H = \frac{ask}{r} \tag{1.12}$$

Entropy thus represents the user-side relevance of the result. In order to assess this, users must first perceive the result (English: "Perception"), understand it, and compare it with their existing knowledge. In the best case, the result fulfills their information need, the information request was thus successful and is completed. However, it may also be that the information need is not yet satisfied, but the user is enabled by capturing the results to better or more precisely formulate their information need. This process is called query refinement. Similarly, the user may mark certain elements of the result as particularly relevant or irrelevant and thus perform an additional action, which also serves to generate a new query that already includes such relevance filters (English: "Relevance Feedback"). In all these cases, after capturing, a refinement or similar additional action takes place, which then triggers a new information request.

From the perspective of the information system, a relevance assessment is also carried out when determining the results. This is also called relevance and arises, for example, from checking the similarity of a data record in relation to the query.

If the data record is similar, it is called a match. Since potentially any number of matches can be found, the information system also performs filtering, sorting, or refinement, thus presenting the optimized and best possible set of matches from the system's side. Figure 1.8 shows these additional tasks of the information system as a complete knowledge/information and data flow.

The considerations and terms listed in this subsection now also help to describe in more detail two fundamental human-computer interaction models for information mediation.

> Information models describe the formal relationships of the individual phases in the information flow.

1.5   Human-Computer Interaction Models

Interaction models are used to specify typical processes between human and machine actors. In software engineering, activity diagrams or use case diagrams are often used for this purpose, which model fairly fine-grained interactions, for example between classes. However, there are also general human-computer interaction patterns that describe entire classes of application systems at an abstract level. The latter category also includes human-computer interaction in the context of knowledge and information. All the scenarios described so far have assumed that users direct their information need to a system. Such scenarios are called human-computer interaction (HCI). Figure 1.8 informally shows the human on the left and the machine on the right.

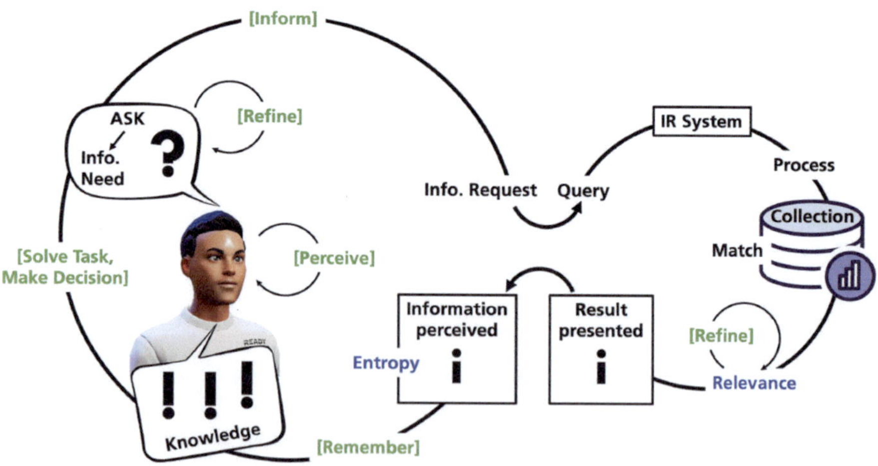

Fig. 1.8 Human-Computer Interaction (HCI) in an information system

> Human-Computer Interaction describes the interaction between humans and machines.

In addition to HCI, there are also models that focus on the interaction between users and consider the machine merely as a supporting platform or component. Such scenarios are referred to as computer-supported cooperative work (CSCW). Figure 1.9 now shows the analogous example in a CSCW scenario.

> Computer-Supported Cooperative Work scenarios use computers to organize the flow of information between people with the help of support systems.

It quickly becomes clear that, for the questioner, nothing seems to change at first. Here, too, the information need must be formulated and a request made. However, in a classic human-to-human interaction, the translation into a query language is likely to be omitted; people simply talk to each other. The data that previously resided in the information system is now replaced by the knowledge of the other person, who formulates an answer from it. To do this, the other person must remember, i.e., retrieve relevant knowledge from memory. Here, too, the terms entropy (from the user's perspective) and relevance (from the other person's perspective) apply, as well as the user's assessment of whether the results fully resolve their ASK. If this is not the case, a refinement and renewed request is initiated.

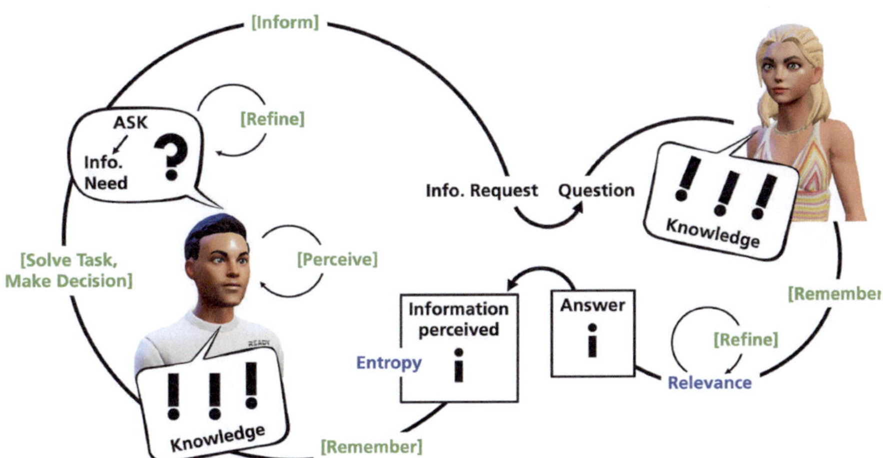

Fig. 1.9 Computer-Supported Cooperative Work (CSCW) in information mediation

In the context of multimedia, a number of additional aspects are emerging in the field of interaction design that are becoming increasingly important. The multitude of existing end devices (e.g., smartphones, tablets, voice control, etc.) as well as technologies from the field of virtual or augmented reality mean that the following aspects must increasingly be taken into account:

- **Social Interaction Design** deals with the interaction between users and their devices as well as with the actions of users among themselves. The focus here is particularly on the dynamics that develop during interpersonal communication. Thus, it is not only about modeling content-related interactions, but above all about the social or even psychological dynamics of the interaction.
- **Emotional Interaction Design** focuses on addressing users not only rationally but also emotionally. This can be achieved through small animations that help users understand things better (Charles Eames: "The details are not the details, the details are the design"), through visualizations that associate a product with positive emotions, or through communication models that leave a positive impression. An excellent example of such emotional interaction design is the language learning app Duolingo. Here, users are initially encouraged by a virtual assistant named "Duo" to practice regularly with the app. However, if this assistant is ignored for a certain period, another assistant suddenly appears with a message such as "Hello, this is Anna. My friend Duo told me that he is very sad because you haven't been in touch. Can I help mediate?". Undoubtedly, this type of communication triggers emotions in the user, perhaps even a slight smile.
- **Sonic Interaction Design** is increasingly used to support aesthetic or emotional content in interactive contexts. The addition of sounds or background noises often triggers an impulse in the user that is also relevant for interactions.

> Interaction design can take place on different levels.

In addition to this classification, Kuhlthau [19] has classified the central characteristics of information-seeking activities, which naturally provide the starting point for information-based interaction with other users or with information systems. Kuhlthau describes these as follows:

- **Starting**: Activities that can be assigned to the initiation of the search
- **Chaining**: Following references (e.g., footnotes or bibliographies, links on the web)
- **Browsing**: a vague search in the area of potential interest
- **Differentiating**: Exploiting differences between already examined material to get an overview of the properties and quality of the material
- **Monitoring**: staying informed about updates from a particular source
- **Extracting**: systematic analysis of a source for relevant material

> There are various forms of information-seeking activities: starting, chaining, browsing, differentiating, monitoring, extracting.

The model by Bates [20] places the human and their knowledge development at the center of the information-seeking process and considers this process in light of insights from the study of humans (anthropology): "Looking at us as a species that exists physically, biologically, socially, emotionally, and spiritually, it is not unreasonable to guess that we absorb perhaps 80% of all our knowledge through simply being aware, being conscious and sentient in our social context and physical environment." Bates identifies various modes that influence search behavior. A core element of her model is that information is acquired not only actively but also passively over time, and that information seeking can be both directed and undirected (see Fig. 1.10).

According to Bates, there are four modes of activities that play a role here:

- **Awareness:** This refers to what we absorb and learn through passive behavior.
- **Monitoring:** Monitoring and browsing are complementary modes. Monitoring is directed and passive, while browsing is undirected and active.
- **Browsing:** There is no specific information need; instead, an exploratory search for new information takes place.
- **Searching:** This refers to a targeted effort to find answers to open questions or to gain an understanding of a new subject.

> Bates' model divides information seeking into the phases: awareness, monitoring, browsing, searching.

It becomes clear that the models of Kuhlthau and Bates do not fundamentally differ from each other, but rather consider different aspects of information activities and thus complement each other in many respects. At this point, it should also be mentioned that interactions do not only take place in the classic way between the user and a computer or smartphone. Wherever we encounter digital devices,

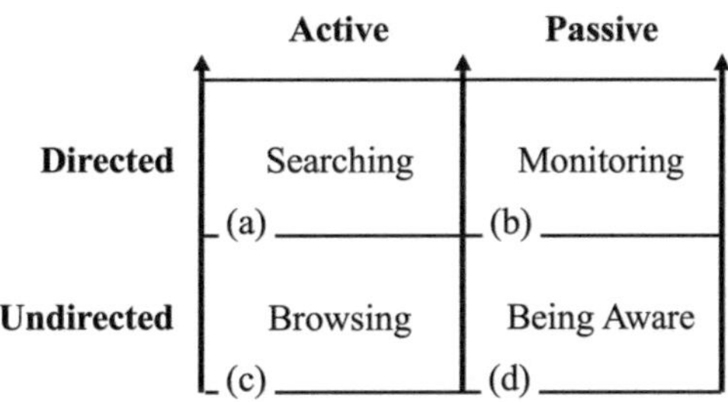

Fig. 1.10 Information Seeking and Searching Model according to Bates [20]

interaction must be regulated and supported. The navigation system in the car, digital signposts, ATMs and parcel pickup stations, all the way to virtual reality—these are all exemplary application areas where interaction design is necessary and where information retrieval plays a central role. Even checking your account balance at an ATM is a classic information retrieval process.

1.6   Multimedia

This subsection introduces the term "multimedia," its forms, and fundamental technical conditions. Furthermore, multimedia features—that is, the representation of the properties of multimedia—are defined and presented in detail. At the beginning, in the Subsections 1.6.1 to 1.6.3, the conceptual properties are first introduced here; Subsection 2 then goes into detail, both in terms of content and technology, on the concrete forms of features, their extraction, and the representation of individual multimedia types. Concrete examples and tools for extraction are also described here.

One of the first mentions of the term "multimedia" dates back to the late 1980s, when the Ford company presented its automobiles at the "1988 Ford New Car Announcement Show" in Detroit in a—quite literally—multimedia presentation (see Fig. 1.11) [21].

Since that time, much has changed, and a good definition of the term "multimedia" is provided by Tay Vaughan [23] as follows: "any combination of text, graphic, art, sound, animation, and video that is delivered by computer or other electronic or digitally manipulated means."

This extremely broad framework immediately demonstrates the importance of the technologies addressed in this book. To put it simply: Multimedia is everywhere!

> Multimedia is everywhere!

1.6.1   Multimedia Collections

The long-term usability of digital media leads to their organization into collections of multimedia objects. The term "collection" is not only used to describe the management of a certain number of digital media objects; a collection is also characterized by the fact that specific descriptive metadata is stored in addition to the actual digital media objects. This additional metadata forms the basis for multimedia information systems (MMIS), in which digital media objects can be searched. Without this metadata, users would have to open and review each individual media object to determine whether it matches their search query. The metadata of a collection is typically used to create a so-called index, which enables targeted content access through efficient searching or exploration (browsing). This is also

Fig. 1.11 Excerpt from the setup of the multimedia presentation for the "1988 Ford New Car Announcement Show." [22]

the foundation for multimedia information retrieval, i.e., targeted content access through search or exploration based on features that are no longer analyzed and managed by humans, but by systems.

1.6.2 Data and Facts

Multimedia has long been an important medium. This leads to ongoing activities in various fields to preserve multimedia objects permanently and in usable condition. For example, the archive of the British Broadcasting Company (BBC) dates back to the early 1890s [24]. Since then, countless projects have been carried out to accommodate ever-emerging digital formats and keep the archive as up-to-date as possible. Currently, the BBC archive comprises 165 petabytes of storage, with approximately 120 terabytes of new material generated each week.

These figures also show that "multimedia" is highly relevant today. And in purely numerical terms, it gets even "worse," because with the advent of social media, the line between classic producers and consumers has blurred, so that now every user creates, distributes, comments on, modifies, or consumes content themselves. Each year, 1.2 trillion digital photos and videos are recorded, now over 85% on smartphones [25]. As early as 2020, Facebook surpassed the mark of one billion active users per month, Instagram reached 500 million daily users of its "Story" feature, and by the end of 2023, 3.4 billion monthly active social media users are expected [26]. The global media and entertainment market has an annual volume of over 2.1 trillion US dollars [27].

Added to this are advances in the level of detail in the digital representation of multimedia objects. For example, current cameras such as the Sony a7R IV35 [28] can produce up to 61 megapixels of resolution, and even smartphones like the Xiaomi Redmi Note 10 Pro [1] push this boundary to 108 megapixels. Multimedia is used in virtually every area of information technology. In the medical sector, X-rays, MRI or CT procedures, as well as real-time recording of ECG or EEG signals, are examples of multimedia objects. The broadcast sector (TV, radio, streaming), the gaming sector (games, consoles), the communications sector (messengers, video conferences, collaboration platforms)—these are just a few examples of multimedia applications. As you can see, "multimedia" is one of the most important and relevant technologies of our time, with which everyone comes into contact directly or indirectly. The sound and structured provision of solutions in this environment is a central task of software development, and a solid grasp of the associated concepts and terms is a key building block for successful projects.

In practice, multimedia objects are typically organized into collections. Only through the organization of individual digital media into a collection do terms such as multimedia information system (MMIS) or multimedia information retrieval (MMIR) become meaningful. The collection thus always defines the framework in which MMIR processes take place and which makes them necessary in the first place. In multimedia collections, multimedia objects of different types can be managed together. For desktop searches, such as Cortana or Spotlight, the entire content of a user's computer can be considered a multimedia collection. In products such as Adobe Lightroom for photo and video production, images and videos are grouped into one or more folders as a collection; on smartphones, apps store their content in app-specific collections; online image databases such as gettyimages present users with pre-made collections; and search engines like Google access an almost infinitely large stock of indexed media. The collection thus defines the content framework in which MMIR takes place and which can then be managed, for example, by an MMIS. The following variants are common:

- **User-defined collections:** Here, users of an application themselves specify which elements are to be considered within the application and thus added to the collection. This is done, for example, by selecting folders containing multimedia objects (see Fig. 1.12).
- **Application-controlled collections:** The application manages the multimedia objects that are, for example, created within the application. A good example of this is the image collections on smartphones, which are usually managed by the respective smartphone app itself and stored in storage areas inaccessible to users. Access to the elements of such collections is only possible via the respective application.

Fig. 1.12 Collection in Adobe Lightroom [29]

- **Provided collections:** Managing digital media can be very labor-intensive. Therefore, collections are often editorially revised and controlled. In this case, a service provider (e.g., Getty Images, Fig. 1.13) takes on the task of reviewing multimedia objects, checking them for certain quality features, enriching them with metadata (to enable effective MMIR, for example), and then making them available to other users as part of a collection. This basic principle is used in a variety of applications. These include:
  - Websites: Every website publishes multimedia objects—even if it's just the displayed text that users see. The set of individual pages, blog posts, images, etc., can be considered a collection, and retrieving a page, blog post, or image is a classic MMIR process.
  - Social Media: Here, the provided collection is created automatically through stored filters, such as the selection of people users follow. From the overall collection of a social media provider, a sub-collection relevant to the current user is determined. Viewed this way, social media is nothing other than an MMIS.
  - TV and Streaming Media: Consuming video in the form of live streams or on-demand video is also based on a provided collection, namely the set of multimedia objects that, for example, Netflix or Amazon makes available to a user. Browsing the collection, selecting, and playing videos are typical MMIR processes. Here, we are dealing with a provided and application-controlled collection, since streaming providers use their own applications to ensure that licensing rights and access restrictions can be enforced.
- **Dynamic collections:** Here, there is no editorial control over new material. Nevertheless, MMIR processes are to be carried out on provided collections.

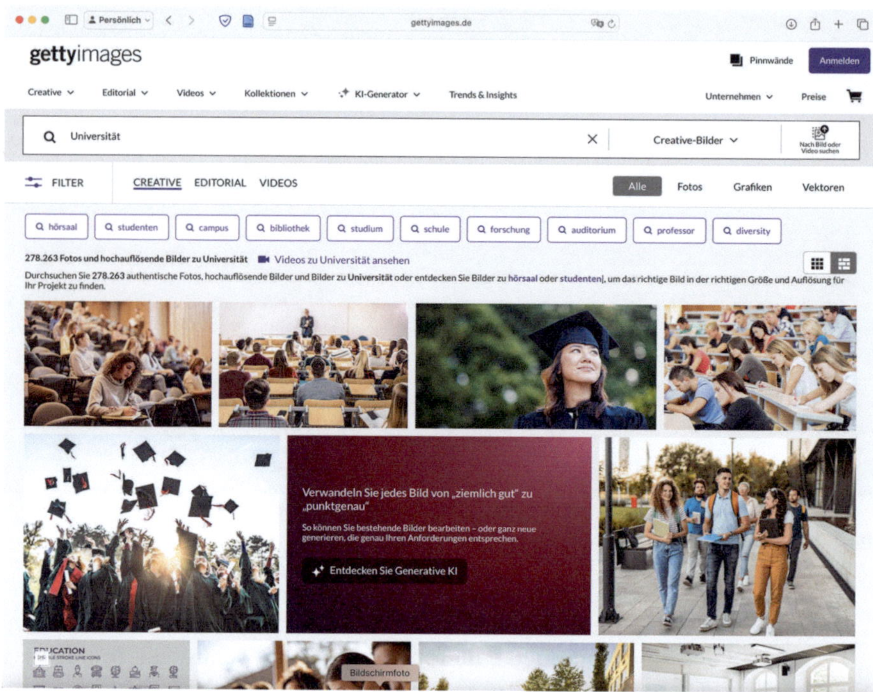

Fig. 1.13 Getty Images collection [30]

An example of this would be the classic Google search, which, based on automatic web crawlers, indexes any number of ever-new websites and adds them to the Google collection.

> Multimedia is organized in collections of all kinds.

The examples in this list once again show how central MMIS and MMIR are in today's world. However, collections not only define the content framework of an MMIR application, they are also a means of capturing, managing, or generating metadata. Simple metadata might be the date a multimedia object was added to a collection or the number of times a multimedia object has been accessed. On this collection-based foundation, MMIR processes such as archiving, moving to another storage medium, or ranking in MMIR can then be managed. Since collections (with the exception of dynamic collections) usually define a manageable scope, it is also possible for users to annotate the elements within a collection with descriptive texts and meta-information. The mere fact that a multimedia object is added to a collection already constitutes meta-information. Therefore, the elements of a collection are typically also multimedia assets and no longer "just" multimedia objects (see definition in the following Subsection 1.6.3).

The following subsections introduce the fundamentals and key terms in the field of "multimedia."

1.6.3 Multimedia Objects

Due to the multitude of technological subfields, formats, and media object types (such as image, audio, video, text), it is important to introduce a set of terms from the outset that will be used throughout the rest of the text.

Multimedia itself is always the representation of a real-world event. The picture or video you take, the text you write—all of these relate to objects or a context of objects and events in the real world. It does not matter whether the text documents a real event or describes fictional events, because even the imagination of an author or the artistic achievement of a painter or composer represents a real process. Even generating artificial images using AI, in the broadest sense, is still an artistic achievement of the person operating the AI. Therefore, the following terms can be introduced and distinguished (see also Fig. 1.14):

- **Real-world object:** This is the original physical scene consisting of objects and events in the real world, an action that is digitally represented and recorded using multimedia.
- **Multimedia content object:** A multimedia representation of the original scene, event, or action in a usually standardized format. For example, this could be an audio file, a video, a photo, or a textual description.
- **Multimedia object:** Within a multimedia content object, various individual digital objects can be identified as representations of objects in the real world. In a text, these could be subsections, sentences, or letters; in an image, the

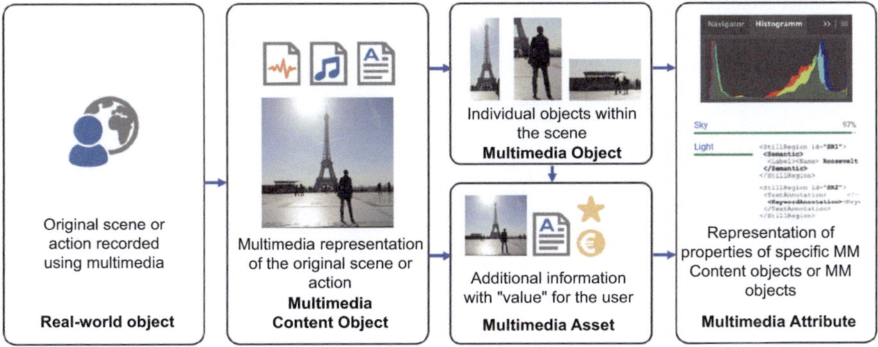

Fig. 1.14 Terms in the context of multimedia objects

people or events depicted; in videos, the individual scenes and also the people or events shown; or in audio, certain time-based information, and much more.

- **Multimedia asset:** Any type of multimedia content object or multimedia object can have "value" for users. This value can be monetary (i.e., if license fees are charged for the use of the multimedia content object, as on streaming platforms), but it can also be virtual, for example, when users add a photo to a "favorite pictures" album or mark a social media post with a "like."
- **Multimedia feature:** These are representations of properties of certain multimedia content objects or multimedia objects that describe information at various levels and in different degrees of detail. Features are often also called "features."

In the world of multimedia terminology, distinguishing between multimedia objects, multimedia assets, and multimedia features is especially important.

In addition to these introductory terms from the field of "multimedia," a basic understanding of the field of "information retrieval" is also necessary to be able to contextualize the following subsections. Even though several subsections are planned for this area later on, the following basic overview should already outline the framework and introduce some key terms (see Fig. 1.15).

"Information retrieval" and the associated "information systems" are typically used to satisfy a user's information need. For example, queries are submitted to a search engine, which then presents results to the user. Everyone is familiar with this behavior from a classic search engine search. However, a number of additional process steps take place in the background: the search engine automatically and continuously generates a so-called index, which enables optimized access to representations of features in order to quickly find the actual results (documents/media objects). This is complicated by the fact that new documents/media objects are constantly being added to the collection of the information system, must be managed, and inserted into the index, and that existing documents/media objects can

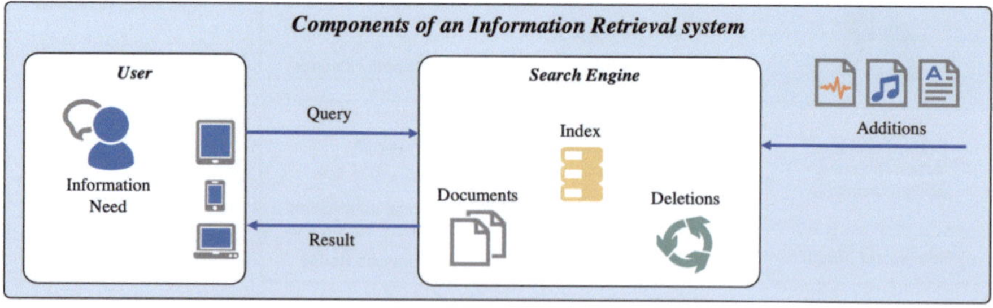

Fig. 1.15   Components of an information retrieval system

also be removed. Classic "information retrieval" focuses on the creation of index structures and the modeling of query and result representations, whereas the term "information system" additionally includes the processes of managing, adding, and deleting multimedia content objects, multimedia objects, and multimedia features. The following abbreviations are used throughout the text:

- MMIR: Multimedia Information Retrieval
- MMIS: Multimedia Information System
- IR: Information Retrieval (as a synonym for MMIR)
- IS: Information System (as a synonym for MMIS)
- MM: Multimedia

The basis for any kind of MMIR and MMIS are MM features and their representation. These will be placed in the overall context in the following subsection.

1.6.4 Multimedia Features

The extraction of MM features from MM content objects and MM objects forms the basis for MMIR and MMIS. Here, different levels of abstraction of features must be distinguished. Using a typical photo, these feature levels can be clearly recognized (see Fig. 1.16). However, it should be noted that other MM content objects, such as video, audio, or text, of course also exhibit other features.

Features can arise at the following levels of abstraction:

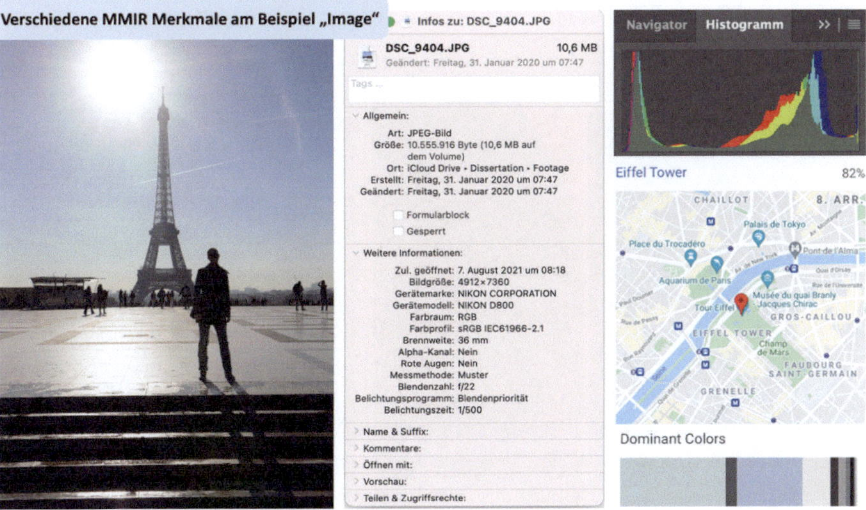

Fig. 1.16 Multimedia feature levels of an image

- **Technical features:** This category includes features that are typically not immediately understandable to humans, such as the binary representation of file formats, control signals, or mathematical calculations such as histograms of photos.
- **General features:** These features describe properties of the MM content object, such as creation date, file size, file name, but also basic metadata of a specific file format (e.g., EXIF data (Exchangeable Image Format) for images), from which information about the camera used or even geoinformation can be extracted.
- **Format-specific features:** Often, generic information can be extracted from a certain type of MM content object. For all images, for example, the dominant colors can always be determined, regardless of what is depicted in the image and with which camera it was taken.
- **Content-based features:** This area is often supported by techniques from the field of pattern recognition. The goal here is to determine, for example, what is depicted in a photo. People, objects, frames, object boundaries, etc., are recognized or calculated.
- **Semantic features:** Here, a meaning is assigned to a content-based feature; for example, a recognized person is described as "Sandra" and a recognized object as the "Eiffel Tower." Spatial relationships such as "standing in front of" can also be introduced. In the text domain, semantic features would be, for example, the automatic recognition of a text's topic.
- **Application-specific features:** These are features that represent a concrete application context and are only meaningful within a specific problem or application domain. Figure 1.17 illustrates this using the example of the DICOM standard (Digital Imaging and Communications in Medicine) [31], which can be used, among other things, for the description of mammography images. This example also makes it clear that features (e.g., feature f10) often have domain-specific semantics. If MMIR is supported or an MMIS is used, it must

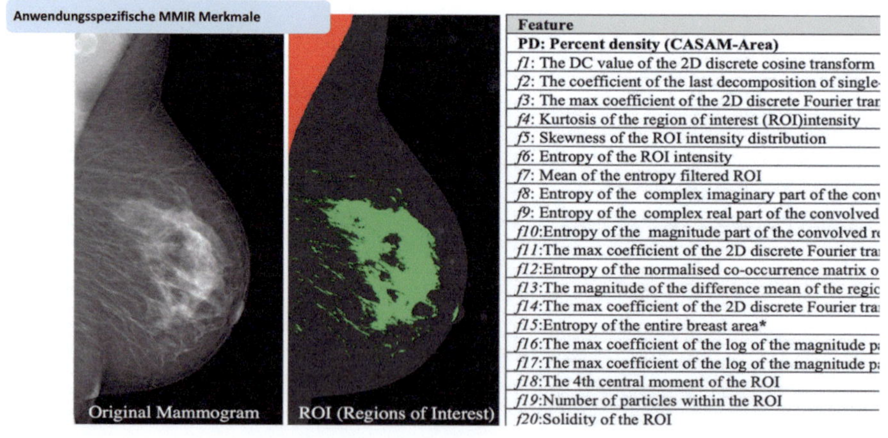

Feature
PD: Percent density (CASAM-Area)
$f1$: The DC value of the 2D discrete cosine transform
$f2$: The coefficient of the last decomposition of single
$f3$: The max coefficient of the 2D discrete Fourier tran
$f4$: Kurtosis of the region of interest (ROI)intensity
$f5$: Skewness of the ROI intensity distribution
$f6$: Entropy of the ROI intensity
$f7$: Mean of the entropy filtered ROI
$f8$: Entropy of the complex imaginary part of the conv
$f9$: Entropy of the complex real part of the convolved
$f10$:Entropy of the magnitude part of the convolved r
$f11$:The max coefficient of the 2D discrete Fourier tra
$f12$:Entropy of the normalised co-occurrence matrix o
$f13$:The magnitude of the difference mean of the regic
$f14$:The max coefficient of the 2D discrete Fourier tra
$f15$:Entropy of the entire breast area*
$f16$:The max coefficient of the log of the magnitude p
$f17$:The max coefficient of the log of the magnitude p
$f18$:The 4th central moment of the ROI
$f19$:Number of particles within the ROI
$f20$:Solidity of the ROI

Fig. 1.17 Application-specific features in the medical context [32]

be ensured that users have a basic understanding of the stored data. Without knowledge of which domain-specific features exist, the MMIR process can become significantly more complicated.

The concrete extraction and representation of features is examined in detail in Chap. 2 below.

> Multimedia features are central elements in the representation of multimedia objects.

1.7   Summary

This chapter introduced the key concepts of information and knowledge understanding as well as their representations. The SECI model provides an initial basic classification. Based on this, the concepts of information and knowledge can be defined (e.g., according to Kuhlen) and the user's information need can be modeled (e.g., according to Belkin). These models then lead to a formalization of informational interaction behavior, for example in the HCI or CSCW context. The concepts of information and knowledge are, of course, also the central component in information retrieval (IR), since IR is concerned with finding and making information accessible. A precise and formal definition of IR can be found in Chap. 3 of this book. In addition, the most important terms from the field of "multimedia" were introduced. In addition to the fundamental and important distinction between multimedia objects, such as multimedia content objects and multimedia assets, the concept of features was also defined. These features in particular form the core of IR applications, as they represent the properties of multimedia objects, which can later be treated as information or knowledge. A main aspect of multimedia information retrieval is therefore working with such features. For this reason, this book will focus intensively on this task in the following chapters.

1.8   Self-Assessment Questions

1.1—Information

What is the relationship between information, knowledge, and action according to Kuhlen?

1.2—Information Need

What is the approach called when the information need is described by providing an example?

1.3—SECI Model

What phases does the SECI model have and what do they stand for?

1.4—Interaction

What is the difference between the HCI and the CSCW model?

1.5—Collections

What basic types of collections do you know?

1.6—Terminology

What is the difference between a multimedia object and a multimedia asset?

1.7—Features

What levels of features do you know?

References

1. Xiaomi Inc. "Redmi Note 10 Pro – The 108MP Voyager." [Online]. Available: https://www.mi.com/global/product/redmi-note-10-pro/overview, Download: 14.12.2020.
2. I. Nonaka and H. Takeuchi, *The Knowledge-Creating Company: How Japanese Companies Create the Dynamics of Innovation*. Oxford University Press, 1995.
3. A. J. Dix, J. E. Finlay, G. D. Abowd, and R. Beale, *Human-Computer Interaction*, 3rd Edition. Harlow, England: Prentice Hall, 2004.
4. I. Greif, *Computer Supported Cooperative Work: Introduction to Distributed Applications*. San Mateo, CA: Morgan Kaufmann Publishers, 1988.
5. R. Kuhlen, *Information: Wissenschaft und Praxis: Die Informationswissenschaftliche Perspektive*. Facultas, 2013.
6. N. Belkin, "Anomalous states of Knowledge as a basis for information retrieval," *Canadian Journal of Information and Library Science*, vol. 5, pp. 133–143, 1980.
7. Facebook.com. "Facebook." [Online]. Available: http://www.facebook.com, Download: 13.10.2020.
8. Google.com. "Google reverse image search." [Online]. Available: https://images.google.com, Download: 14.10.2020.
9. A. M. Keller and F. Korn, "Query by example," *ACM Computing Surveys (CSUR)*, vol. 21, no. 1, pp. 29–58, 1989.
10. J. Harrington, "Introduction to sql," in Dec. 2010, pp. 65–74, ISBN: 9780123756978. DOI: 10.1016/B978-0-12-375697-8.50003-0.
11. F. Inc., *Graphql: A query language for your api*, https://graphql.org/, Letztes Zugriffsdatum..
12. W3C. "SPARQL Query Language for RDF." [Online]. Available: https://www.w3.org/TR/sparql11-overview/, Download: 14.12.2021.
13. Devart. "General review of microsoft sql server management studio (ssms)," Devart. [Online]. Available: https://blog.devart.com/general-review-of-microsoft-sql-server-management-studio-ssms.html, Download: 11.08.2021.
14. Internet Engineering Task Force (IETF). "The JavaScript Object Notation (JSON) Data Interchange Format." [Online]. Available: https://datatracker.ietf.org/doc/html/rfc7159, Download: 01.02.2021.

15. W3C. "Simple Object Access Protocol Specification," W3.org. [Online]. Available: https://www.w3.org/TR/soap/, Download: 11.07.2021.
16. Wikipedia. "Restful services, representational state transfer," Wikipedia. [Online]. Available: https://en.wikipedia.org/wiki/Representational_state_transfer, Download: 11.08.2021.
17. D. Jurafsky and J. H. Martin, *Natural Language Processing*, 3rd Edition. Harlow, England: Pearson, 2020.
18. Open.AI, *Chat gpt*, https://chat.openai.com/, 2023.
19. C. C. Kuhlthau, "Seeking meaning: A process approach to library and information services," *Libraries Unlimited*, 2004.
20. M. Bates, "Toward an integrated model of information seeking and searching," *New Review of Information Behaviour Research*, vol. 3, pp. 1–15, 2002.
21. K. Woolford. "New photography: A perverse confusion between the real and the live." [Online]. Available: https://www.researchgate.net/publication/336030256_New_Photography_A_Perverse_Confusion_Between_the_Real_and_the_Live, Download: 01.10.2020.
22. W. g. Moproducer. "Ford show 1987.jpg." [Online]. Available: https://commons.wikimedia.org/wiki/File:Ford_Show_1987.jpg?uselang=de.
23. T. Vaughan, *Multimedia: making it work*. Berkeley, Calif: Osborne McGraw-Hill, 1993, 477 pp., ISBN: 978-0-07-881869-1.
24. A. Williams. "Safeguarding the bbc's archive." [Online]. Available: https://www.bbc.co.uk/blogs/bbcinternet/2010/08/safeguarding_the_bbcs_archive.html, Download: 23.10.2020.
25. M. Nudelman. "Smartphones cause a photography boom." [Online]. Available: http://www.businessinsider.com/12-trillion-photos-to-be-taken-in-2017-thanks-to-smartphones-chart-2017-8, Download: 22.04.2021.
26. J. Clement. "Social Media – Statistics & Facts." [Online]. Available: https://www.statista.com/topics/1164/social-networks/, Download: 23.10.2020.
27. S. GmbH. "Value of the global entertainment and media market from 2011 to 2024." [Online]. Available: https://www.statista.com/statistics/237749/value-of-the-global-entertainment-and-media-market/, Download: 02.08.2020.
28. Sony Electronics Inc. "a7R IV 35 mm full-frame camera with 61.0 MP. "[Online]. Available: https://www.sony.com/electronics/interchangeable-lens-cameras/ilce-7rm4, Download: 03.10.2020.
29. Adobe Systems Software Ireland Ltd. "Creative journeys start here." [Online]. Available: http://www.adobe.com.
30. G. Images. "Getty images." [Online]. Available: https://www.gettyimages.de.
31. DicomLookup.com. "Dicom lookup." [Online]. Available: http://dicomlookup.com/.
32. I. Maglogiannis, C. Doukas, G. Kormentzas, and T. Pliakas, "Wavelet-based compression with roi coding support for mobile access to dicom images over heterogeneous radio networks," *Information Technology in Biomedicine, IEEE Transactions on*, vol. 13, pp. 458–466, 2009. 10.1109/TITB.2008.903527.

Features and Feature Representation

2

This chapter deals with the concrete manifestation and representation of multimedia features. Since these are highly dependent on the type of multimedia object, this chapter is divided into four major subsections. Subsection 2.1 first describes textual features, their representation, and encoding. Since textual representations also form an important basis for the other multimedia features, all concepts introduced here apply analogously to the other multimedia types. Subsection 2.2 then contains information about images, their technical representation, and feature characteristics. In Subsection 2.3, audio features as well as the physical/technical fundamentals of audio encoding are described, and in Subsection 2.4, the topic of video is addressed. In all areas, the content-related/technical fundamentals are first introduced, digital encoding formats and standards are specified, and mechanisms and tools for feature extraction are listed. For didactic reasons, this subsection also considers the history of the individual formats, in order to enable a well-founded and deep understanding of the interrelationships and technologies [7].

2.1 Text Features

Text features represent an important pillar of feature representation. On the one hand, they are used to represent the content of text documents; on the other hand, they can also be used for the textual representation of arbitrary features of other multimedia types. Therefore, text features are introduced at the beginning as a cross-sectional function for all multimedia types and, naturally, explained using the example of text documents. We thus begin with the fundamental components of a text document, namely the content, and how this content can be digitally represented or encoded [8]. Initially, we assume that the content consists of "plain" text, i.e., it does not yet contain any formatting, embedded multimedia objects, or similar elements. Examples of how to handle formatted text or special document

© The Author(s), under exclusive license to Springer-Verlag GmbH, DE, part of Springer Nature 2026

M. Hemmje and S. Wagenpfeil, *Multimedia Information Retrieval*,
https://doi.org/10.1007/978-3-662-73310-3_2

formats can be found at the end of the chapter under 2.5.1. So, let us first assume that the content of a document consists of a sequence of elementary text units, such as letters, digits, punctuation marks, and other characters. Ultimately, these must be encoded in bits and bytes. The specification for this is called character encoding.

At this point, it should be noted that—regardless of the type of document—a computer must encode every form of information in bits and bytes in order to store, display, or distribute it. In the digital world, there are only zeros and ones to describe information. Thus, for example, every letter, every number, every pixel of an image, etc., must be encoded as a sequence of zeros and ones. For example, the letter "A" in the US-ASCII format can be represented by the decimal number 65 or the binary sequence 1000001. The letter "B" would be the decimal number 66 or the binary sequence 1000010. Binary sequences can be translated into decimal numbers by considering each position (i.e., each bit) as a power of two from right to left and adding them up [8]. The decimal number 66 is thus derived from the 7-digit binary sequence 1000010 as follows:
$$2^6 \cdot 1 + 2^5 \cdot 0 + 2^4 \cdot 0 + 2^3 \cdot 0 + 2^2 \cdot 0 + 2^1 \cdot 1 + 2^0 \cdot 0 = 64 + 0 + 0 + 0 + 0 + 2 + 0 = 66$$
. In addition to binary and decimal representation, hexadecimal representation is also frequently used. Hexadecimal numbers consist of digits that can each take on 16 different values (0–9, A–F). Each of these digits can be assigned both a decimal and a binary value (see Table 2.1).

> Characters of any kind are represented by binary sequences.

Table 2.1 Decimal, hexadecimal, and binary representation

Decimal	Hexadecimal	Binary
0	0	0000
1	1	0001
2	2	0010
3	3	0011
4	4	0100
5	5	0101
6	6	0110
7	7	0111
8	8	1000
9	9	1001
10	A	1010
11	B	1011
12	C	1100
13	D	1101
14	E	1110
15	F	1111

To convert binary numbers to hexadecimal numbers, you always start from the 4-digit binary representation, i.e., every 4 bits encode one hexadecimal digit. Missing bits are supplemented with leading zeros. In the example above, this means that the binary sequence 1000010 would be supplemented with a leading zero and then divided into two blocks of four: 0100 and 0010. The first block yields the hexadecimal digit 4, the second yields 2, and thus the hexadecimal number 42. As you read through the following subsections, keep in mind that all features or their representations must be encoded in binary in this way in order to be processed on digital devices at all.

One of the best-known character encodings for text is US-ASCII, standardized as ANSI X3.4 (1968) and as the ISO 646-US variant of ISO 646 (1972) [9]. The character set of US-ASCII (Fig. 2.1) is only sufficient to encode Latin and US English texts as well as texts in a few other languages. For German, Danish, and fourteen other languages, ISO 646 therefore provides national variants [10], which replace the US-ASCII characters "#" (hash), "$" (dollar), " (at), "[" (left square bracket), "\ " (backslash), "]" (right square bracket), "^" (caret), "'" (grave), "{" (left curly bracket), "|" (vertical bar), "}" (right curly bracket), and "~" (tilde) with national characters. The German variant ISO 646-DE (Fig. 2.2), for example, makes the substitutions listed in Table 2.2 [11].

For this reason, a sender "Technische Universität München, Arcisstraße 21" may still occasionally appear in American email programs as "Technische Universitt Mnchen, Arcisstraße 21"; that is, the address originally encoded with ISO 646-DE was output using ISO 646-US.

Character sets standardize text encoding.

In the mid-1980s, the European Computer Manufacturer's Association (EC-MA) extended the US-ASCII character set into a family of character sets capable of encoding alphabetic writing systems. The character set family, now codified by

hex	0	1	2	3	4	5	6	7	8	9	A	B	C	D	E	F
2_		!	"	#	$	%	&	'	(	)	*	+	,	−	.	/
3_	0	1	2	3	4	5	6	7	8	9	:	;	<	=	>	?
4_	@	A	B	C	D	E	F	G	H	I	J	K	L	M	N	O
5_	P	Q	R	S	T	U	V	W	X	Y	Z	[	\	]	^	_
6_	`	a	b	c	d	e	f	g	h	i	j	k	l	m	n	o
7_	p	q	r	s	t	u	v	w	x	y	z	{	|	}	~	

Fig. 2.1 The character set ISO 646-US (US-ASCII) [12]

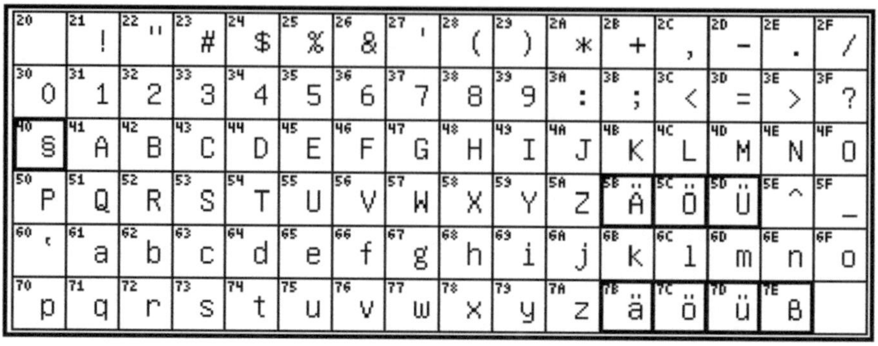

20	21 !	22 "	23 #	24 $	25 %	26 &	27 '	28 (	29)	2A *	2B +	2C ,	2D −	2E .	2F /
30 0	31 1	32 2	33 3	34 4	35 5	36 6	37 7	38 8	39 9	3A :	3B ;	3C <	3D =	3E >	3F ?
40 §	41 A	42 B	43 C	44 D	45 E	46 F	47 G	48 H	49 I	4A J	4B K	4C L	4D M	4E N	4F O
50 P	51 Q	52 R	53 S	54 T	55 U	56 V	57 W	58 X	59 Y	5A Z	5B Ä	5C Ö	5D Ü	5E ^	5F _
60 `	61 a	62 b	63 c	64 d	65 e	66 f	67 g	68 h	69 i	6A j	6B k	6C l	6D m	6E n	6F o
70 p	71 q	72 r	73 s	74 t	75 u	76 v	77 w	78 x	79 y	7A z	7B ä	7C ö	7D ü	7E ß	7F

Fig. 2.2 The German variant ISO 646-DE of ISO 646 [12]

Table 2.2 The substitutions in the German variant of ISO 646

ISO 646-US	ISO 646-DE
[	Ä
\	Ö
]	Ü
{	ä
|	ö
}	ü
~	ß
@	§

ISO under the name ISO 8859, consists of the character sets ISO 8859-1 through ISO 8859-15. Each ISO 8859-X character set includes the 128 US-ASCII characters and adds another 128 characters. For most Western European languages, such as German, Danish, or French, ISO 8859-1, also known as ISO Latin-1 (Fig. 2.3), is relevant. ISO 8859-7 covers the Greek alphabet, and ISO-8859-15 essentially replaces the international currency symbol with the euro sign.

In the early 1990s, Unicode and ISO/IEC 10646 were introduced [13]. The goal of these two character encodings was universality, i.e., the ability to uniquely encode texts from all languages of the world. With Unicode version 10, a unification of the two formats Unicode and ISO/IEC 10646 was achieved in 2017. The current Unicode 14.0 version encodes a total of 144,697 characters, covering all modern and many classical languages. Unicode is continuously being expanded to include further historically significant and currently relevant characters and languages. A good example of this is the inclusion of emoticons in the Unicode standard. The following Fig. 2.4 shows some examples of this and their Unicode character codes [11].

A0	A1 ¡	A2 ¢	A3 £	A4 ¤	A5 ¥	A6 ¦	A7 §	A8 ¨	A9 ©	AA ª	AB «	AC ¬	AD -	AE ®	AF ¯
B0 °	B1 ±	B2 ²	B3 ³	B4 ´	B5 µ	B6 ¶	B7 ·	B8 ¸	B9 ¹	BA º	BB »	BC ¼	BD ½	BE ¾	BF ¿
C0 À	C1 Á	C2 Â	C3 Ã	C4 Ä	C5 Å	C6 Æ	C7 Ç	C8 È	C9 É	CA Ê	CB Ë	CC Ì	CD Í	CE Î	CF Ï
D0 Ð	D1 Ñ	D2 Ò	D3 Ó	D4 Ô	D5 Õ	D6 Ö	D7 ×	D8 Ø	D9 Ù	DA Ú	DB Û	DC Ü	DD Ý	DE Þ	DF ß
E0 à	E1 á	E2 â	E3 ã	E4 ä	E5 å	E6 æ	E7 ç	E8 è	E9 é	EA ê	EB ë	EC ì	ED í	EE î	EF ï
F0 ð	F1 ñ	F2 ò	F3 ó	F4 ô	F5 õ	F6 ö	F7 ÷	F8 ø	F9 ù	FA ú	FB û	FC ü	FD ý	FE þ	FF ÿ

Fig. 2.3 The character set ISO 8859-1 (ISO Latin-1) [12]

Unicode character		Description	Official designation
U+1F600 (128512)	😀	Grinning face	GRINNING FACE
U+1F601 (128513)	😁	Grinning face with smiling eyes	GRINNING FACE WITH SMILING EYES
U+1F602 (128514)	😂	Face with tears of joy	FACE WITH TEARS OF JOY
U+1F603 (128515)	😃	Face with open mouth	SMILING FACE WITH OPEN MOUTH
U+1F604 (128516)	😄	Face with open mouth and smiling eyes	Smiling face with open mouth and smiling eyes
U+1F605 (128517)	😅	Face with open mouth and a bead of sweat	SMILING FACE WITH OPEN MOUTH AND COLD SWEAT
U+1F606 (128518)	😆	Face with open mouth and tightly closed eyes	SMILING FACE WITH OPEN MOUTH AND CLOSED EYES
U+1F607 (128519)	😇	Angel face	SMILING FACE WITH HALO

Fig. 2.4 Unicode extensions for emoticons

> Unicode is currently the most common and universal standard for character encoding.

So far, the focus has mainly been on the characters and their representations within a code table. However, there are several other equally important components of character encodings, which will be examined in more detail below. This results in a multi-level encoding model [8], which includes an abstract character set, an encoding format, a code table, an encoding scheme, and a transfer syntax. The individual components of such an encoding model correspond to increasingly concrete representations of characters, from an abstract concept to bit patterns. Thus, there is always a mapping between an abstract representation level and the next, more concrete representation level. To move from a character position in a code table directly to its encoding in an encoding scheme across several levels, the various mappings can be performed in sequence, resulting in a character map.

2.1.1 Abstract Character Set

The highest, most abstract level of the encoding model [8] is the abstract character set, which consists of a set of abstract characters. An abstract character is a unit of information used for the representation, organization, or control of text, such as letters, digits, punctuation marks, accents, graphic symbols, ideographic characters, spaces, tabs, line feed and carriage return, control codes for selection markers (e.g., "Start of Selected Area" and "End of Selected Area," etc.). The character set of US-ASCII, for example, consists of 33 control characters and 95 printable characters.

However, an abstract character set does not yet include any ordering or numbering of the characters it contains. This aspect is only added at the next stage of the encoding model in the form of a code table. Thus, abstract character sets initially contain all possible elements needed to encode all texts of a particular language or language family. Out of the vision to encode all living and all historically significant languages with a single character set, the most important character set today, Unicode, was created [14]. For the visual presentation of the individual components of an abstract character set, abstract graphic units called glyphs are used. Glyphs correspond at the presentation level to the characters at the encoding level.

Figure 2.5 contains graphic manifestations of the glyph for the letter "A," which are often also called glyph images. Glyph images that belong together in terms of design, size, weight, and other characteristics form a graphic character set of glyph images (font). Such fonts can be installed on the respective operating systems and used in the corresponding applications. At this point, it is important not to restrict oneself to a somewhat trivial view based on the German or English language area. The languages we are typically familiar with usually have a fairly simple

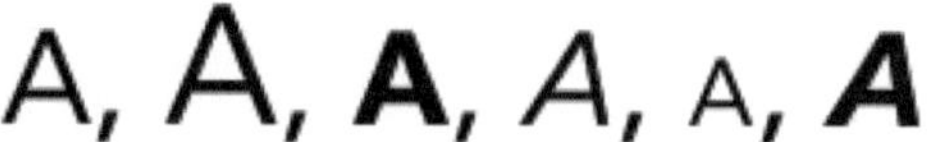

Fig. 2.5 Graphic manifestations of the glyph for the letter "A"

relationship between glyph and abstract character. However, there are languages in which abstract characters are represented by several glyphs, or in which individual glyphs play a role for several abstract characters. Think here of Japanese, Chinese, or even Arabic script characters. In addition to the original language, graphic, optical, or semantic conditions can also play a role. Gutenberg, for example, used glyphs of different widths for the same letter or abbreviating symbols for words or syllables to create a uniform right margin. Here, formatting provides the context for the choice of a glyph. In mathematical expressions, semantics provide a context for the selection of glyphs. For example, a long or short arrow can be chosen depending on whether it is the type specification of a function ($f : A \rightarrow B$) or a production in a grammar ($S \longrightarrow [S]$). In many cases, the elementary text units that should be included as abstract characters in a character set are not obvious. Requirements and conflicts can arise both from the various languages whose texts are to be encoded and from the various processing functions to be applied to the texts. For example, for the display of a German text in printed form, distinguishing between uppercase and lowercase letters in the character set is very convenient. For sorting functions or comparisons, however, it presents a certain hurdle. Defining an abstract character set therefore requires careful consideration of various alternatives, taking into account languages, scripts, and processing functions.

In the case of Unicode, a number of design principles were therefore formulated [13], which operate at various levels and are intended to simplify, standardize, and implement the processing of text data. The first two important design principles of Unicode are already found at the character encoding level:

- Unicode encodes characters, not glyphs.
- Unicode encodes plain text, not formatting information.

2.1.2 Code Table

As already described, the code table provides a mapping rule that assigns a code position to the abstract characters in the character set. Code positions are always natural numbers greater than or equal to zero. The set of possible code positions is also called the code space [8] and is usually given in hexadecimal notation. Each position in the code space thus describes an abstract character and can be uniquely identified by a hexadecimal number or the corresponding bit pattern (see also Table 2.1). In code tables, certain code positions are often left unused and reserved for control purposes.

For example, positions 0 to 1F and position 7F in US-ASCII are reserved for control purposes, as are positions 80 to 9F in ISO 8859-X [13]. The Unicode code space (see Fig. 2.6) contains gaps at the 2048 (decimal) surrogate positions from D800 to DFFF and at positions FFFE and FFFF. These positions are reserved for special functions, which will be discussed in more detail under the aspects of encoding format and encoding scheme. No Unicode characters will be assigned to them in the future either. The 6,400 Unicode code positions from E000 to F8FF,

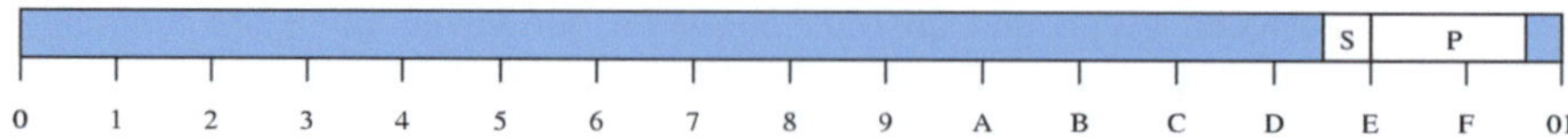

Fig. 2.6 The Unicode code space with surrogate positions (S) and private area (P)

just under 10% of the code space from 0 to FFFF, are reserved for private use. These positions are released for logos, icons, or for characters needed in special notations (music, dance) outside the Unicode character set. The code positions in this area will also be reserved for such private use in future Unicode extensions.

For each character encoded in its code table, Unicode includes the following data (see also Fig. 2.4): the code position of the character, a typical glyph image, the name of the character, and semantic information, such as a textual description. At the code table level, further Unicode design principles can also be found:

- Unicode represents characters that occur in different alphabets but look similar only once (unification).
- Convertibility between established standards: A one-to-one mapping from an established standard to Unicode should be possible. Character positions from a single internationally widespread character set that, according to the unification design principle, should actually be identified with each other, nevertheless receive separate Unicode code positions.
- Unicode defines semantic properties for characters.
- Each base character can be paired with any number of combining characters simultaneously.
- Characterization of equivalent encodings: For the various encodings of a base character with combining characters or of a primary character and a character included in the character set for compatibility reasons, a normalized encoding can be found.

Now that the code table has been defined, the individual positions of the code space can be represented in binary. For this, a so-called encoding format is used.

2.1.3 Encoding Format

An encoding format for a code table defines bit representations for the code position. For this purpose, it uses a code unit, which usually consists of eight or sixteen bits. An encoding format then maps the positions of a code space to sequences of code units and thus to bit patterns. If each code position in a code table is mapped to the same number of code units, we speak of a fixed-length encoding format; otherwise, the encoding format is variable-length. A fixed-length encoding format results from the binary representation of code positions. The number of code units required is determined by the size of the code space and the number of bits in a code unit. Table 2.3 provides an overview of several different encoding formats.

Table 2.3 Encoding formats for various encodings

Encoding/Encoding Format	Code Space/Code Unit/Length
US-ASCII	0-7F/1 Byte/fixed (1)
ISO 8859-X	0-FF/1 Byte/fixed (1)
ISO/IEC 10646, Plane 0/UCS-2	0-FFFF/2 Byte/fixed (1)
ISO/IEC 10646/UCS-4	0-7FFFFFFF/4 Byte/fixed (1)
Unicode/UTF-8	0-10FFFF/1 Byte/variable (1-4)
Unicode/UTF-16	0-10FFFF/2 Byte/variable (1-2)

In principle, an encoding can define multiple encoding formats. Unicode, for example, defines only two encoding formats, namely UTF-8 and UTF-16, where UTF stands for "Unicode Transfer Format." UTF-8 has a code unit of 8 bits in length and represents code positions with one to four code units. UTF-16 has a code unit of 16 bits in length and represents code positions with one to two code units. Originally, Unicode was designed for a code space from 0 to FFFF, so 16 bits would have been sufficient to represent code positions. In fact, this code space would have sufficed at least up to version 3.0. However, when it became clear that the code space would not be sufficient for all historically significant languages and alphabets, the concept of surrogate code positions was introduced and the Unicode code space was extended up to position 10FFFF.

To encode characters outside the original code space in UTF-16, the code number of the character (here called U) is first reduced by 65536 (10000hex = size of the standardized and original code table), resulting in a 20-bit number U' in the range from 00000hex to FFFFFhex. This is then split into two blocks of 10 bits each:

- The first block (i.e., the 10 most significant bits of code U') is prefixed with the bit sequence 110110; the resulting 16-bit word of two bytes is called the high surrogate.
- The second block (i.e., the 10 least significant bits of code U') is prefixed with the bit sequence 110111; the resulting 16-bit word of two bytes is called the low surrogate.

The following code ranges are specifically reserved for such surrogates, i.e., UTF-16 surrogate characters, and therefore do not contain independent characters:

- from U+D800 to U+DBFF ($2^1 0 = 1024$ high surrogates)
- from U+DC00 to U+DFFF ($2^1 0 = 1024$ low surrogates).

When converting UTF-16 encoded character strings into UTF-8 byte sequences, it is important to note that pairs of high and low surrogates must first be recombined into a single Unicode character code before it can be converted into a UTF-8 byte sequence. These mechanisms also ensure that, despite the enormous expansion of the code table, efficiency as a key design principle is maintained. For each of the two Unicode encoding formats, UTF-8 and UTF-16, it is possible to find the start of the corresponding code position representation from any code unit in a stream of position representations. In the case of UTF-16, at most one code unit needs to be stepped back, and in the case of UTF-8, at most three positions.

After this rather detailed information on character encoding and especially the important Unicode standard, the physical storage of documents should now be clarified. You have learned how individual characters of a text can be systematically and securely mapped to binary sequences, character sets, encodings, and code tables. So far, we have always spoken of "plain" text for this purpose (with a few control characters excepted). However, in practice, of course, many other formats are used, which will be briefly explained in the next subsection.

> When exchanging textual information, encoding formats must be taken into account.

2.1.4 Text Formats and Documents

Most text formats today are no longer human-readable and require special applications to display them in a comprehensible form. Fortunately, these applications are standardized, and everyone can, of course, open Word or PDF documents. However, it must be clear that the machine-readable format differs significantly from the displayed format, as the pure text information is enriched here with metadata about formatting, links, or embedded multimedia objects. The semantics of formatting also represent a multimedia feature, since, for example, bold text may have a more important meaning than normally formatted text. Examples of common formats include:

- PDF: The Portable Document Format [15] was specified by Adobe and preserves fonts and layout information. It has become the de facto standard for document exchange [15].
- DOC/DOCX: This format originated in Microsoft Word and is now based on a standardized XML schema, Open Office XML [16]. Documents are typically edited with Microsoft Word or Open Office, but due to the XML structure, they can also be edited in many other applications.
- TXT: The TXT format typically contains plain text files in which formatting plays no role, no layout options are available, and only the pure textual information is conveyed. This format is also often used in technical contexts, for example, to describe lists in comma-separated form (CSV file, "Comma Separated Value") and thus enable simple, standardized data exchange.
- HTML: The Hypertext Markup Language [17] was developed to provide content for the Internet and to allow simple formatting and linking. Here, a plain

text document is enriched with additional markup, e.g., <h1> </h1>, which is then interpreted and formatted accordingly by the browser.

- XML: The eXtensible Markup Language [18] is a format that enables the automated exchange of information and the standardized description of domain-specific documents. XML documents are well-formed (i.e., every opening tag is also closed) and verifiable (i.e., their structure can be validated against a schema).
- JSON: The JavaScript Object Notation format [3] was introduced as a supplement to XML and is intended to optimize the technical exchange of information. Here, the possibility of validity checking is omitted, and the XML tags are replaced by a syntactic structure of various brackets. This results in significantly leaner documents for data exchange.

The major challenge now is to separate the content of text documents from the structural information and subject it to analysis. Such analyses are based on the principle of indexing, i.e., the extraction of features such as textual terms or other textual properties that are relevant for a particular document.

2.1.5 Analysis and Indexing of Documents

One of the central tasks in the automatic processing of documents and also in information retrieval is the content analysis of texts. This involves considering a document not just as a sequence of bits and bytes, but as a content-based representation. Such representations contain the typical features of all documents in a collection and are called an index. The index can be used to optimize access, perform relevance calculations, sort results, or extract information about a collection or individual documents. In this subsection, we will first focus on the basic techniques for creating an index, and then apply these to documents and other multimedia types [19, 20].

Documents (and, of course, all other multimedia content objects) are indexed to enable users to quickly search and access them. Imagine a 400-page text document and a collection of thousands of such documents. A user is now searching for documents in which the term "Einstein" appears. Without an index, your only option is to open and read all documents. Only after processing the last document can you determine a set of matches and present them to the user—a process that can take a long time. An index is a structure that allows us to perform a preliminary analysis of the document in order to optimize the retrieval process later. For example, all nouns are extracted from all documents and stored centrally. The search for "Einstein" would then first query this central location and immediately find a list of documents in which the noun "Einstein" appears—also called an inverted index. In addition, the frequency of the word "Einstein" in the document would be stored here, allowing the results to be sorted by relevance. To enable such optimizations, four steps and techniques are necessary:

- **Merging text units:** Here, the contents of the documents are extracted, formatting is eliminated, and control characters are interpreted.
- **Tokenization:** The content of the document is broken down into individual distinguishable units.
- **Linguistic analysis:** Words are reduced to their stem, and synonyms are taken into account.
- **Inversion of the index:** The identified index terms are transferred into a more usable structure and reference the original documents.

2.1.5.1 Merging Text Units

In this first step, care is taken to identify the relevant information in the documents. The structure and organization of the document collection play a major role here. If your documents, for example, consist of several chapters, you may want to return not the entire document but the relevant chapter as a result. This also affects the search process itself: imagine a user enters two words as a search query—one is found in the first chapter, the other in the last chapter of a document, but no chapter contains both words. If you index based on the entire document, you would have a match. However, a chapter-based index would yield no complete match, since only part of the query could be answered in each case. Such decisions are usually application- or media-specific.

For further processing, the character set (see Subsection 2.1.1), the encoding format (see Subsection 2.1.3), and also the document format (see Subsection 2.1.4) now play an important role. While character set and encoding format are more technical framework conditions, the document format may have a dual function: on the one hand, the document format specifies the storage structure, and on the other hand, it contains a number of control characters or formatting that can provide important information to the indexer. For example, if a text in an HTML document is marked with an <h1> tag and thus formatted as a heading, the indexer could use this information to assign higher relevance to the corresponding text.

Once these decisions have been made, the extracted content of the text units (i.e., the subsections, chapters, pages, documents of the collection) can be further processed. Techniques such as vectorization and tokenization play an important role here.

2.1.5.2 Vectorization

Vectorization is one of the simplest ways to transform texts into mathematically more tractable structures [21]. As an example, consider the following text extracted from a document: "There is a person standing on the street. The person wears a hat and a blue jacket. The jacket and the hat are wet."

First, this text is split into individual sentences:

- S1: There is a person standing on the street
- S2: The person wears a hat and a blue jacket
- S3: The jacket and the hat are wet

Table 2.4 A simple vectorization table

Sentence	There	Is	A	Person	Standing	On	The	Street	Wears	Hat	And	Blue	Jacket	Are	wet
S1	1	1	1	1	1	1	1	1	0	0	0	0	0	0	0
S2	0	0	2	1	0	0	1	0	1	1	1	1	1	0	0
S3	0	0	0	0	0	0	2	0	0	1	1	0	1	1	1

During the extraction process, parsing the document also produces additional information: the vocabulary V. This represents the set of all words in the document and, in our example, would be the following list: $V = \{there, is, a, person, standing, on, the, street, wears, hat, and, blue, jacket, are, wet\}$

In principle, vectorization now means that for each sentence, the frequency of the words in the vocabulary is determined. That is, the individual sentences are essentially mapped over the list of vocabulary terms, and their frequencies are recorded:

- S1: [1,1,1,1,1,1,1,1,0,0,0,0,0,0,0]
- S2: [0,0,2,1,0,0,1,0,1,1,1,1,1,0,0]
- S3: [0,0,0,0,0,0,2,0,0,1,1,0,1,1,1]

It is important to note here that these word vectors are all the same length—namely, the length of the vocabulary vector. This is very helpful for later calculations. This results in a complete **vectorization table** for the document (see Table 2.4).

Based on such a vectorization table, a whole range of calculations can now be performed. For example, the frequency of words in certain subsections or in the entire document can be calculated. Therefore, vectorization tables are sometimes calculated for the entire document, but often also for individual sections. Many statistical methods for determining relevance are also based on vectorization tables. For example, one could argue that the "most relevant" words in a text are those that occur most frequently, i.e., those with the highest sum in a vectorization table. However, it should be noted that in this case, you would most likely get "the," "a," "is," etc., as highly relevant words, whereas the name of the author of a text, which may only be mentioned once on the title page, would be classified as "irrelevant." As you can see, there is still plenty of room for optimization. In the simplest case, a good optimization can already be achieved using so-called stop word lists. These contain those words from the vocabulary that are classified as "irrelevant" within a given relevance categorization. In the example above, the vocabulary terms "is," "a," "the," "and," "are" could be such stop words. A major advantage of vectorization is its simple implementation, as shown in Listing 2.1.

```java
1  // Read file
2  RandomAccessFile inputDoc = new RandomAccessFile("Simpletext.
       txt", "r");
3  String content = "";
4  String line = "";
5  while ((line = inputDoc.readLine()) != null) {
6      content += line;
7  }
8
9  // Detect sentences
10 String[] sentences = content.split(".");
11
12 // Detect vocabulary
13 Vector<String> voc = new Vector<String>();
14 String[] words = content.split("␣");
15 for (String word : words) {
16     // Satzzeichen usw. entfernen
17     word = word.replace(".", "␣");
18     word = word.replace(",", "␣");
19     word = word.trim();
20     if (!voc.contains(word)) voc.add(word);
21 }
22
23 // Vectorization
24 int[][] vec = new int[sentences.length][voc.length];
25 for (int sentence_idx = 0; sentence_idx < sentences.length;
       sen-tence_idx ++) {
26     String sentence = sentences[sentence_idx];
27     String sWords = sentence.split("␣");
28     for (String word : sWords) {
29         word = word.trim();
30         if (voc.contains(word)) {
31             int voc_index = voc.indexOf(word);
32
33             // Eintrag in Vektortabelle
34             vec[sentence_idx][voc_index] = vec[sentence_idx][
                   voc_index] + 1;
35         }
36     }
37 }
```

Listing 2.1 Vectorization in Java

Vectorization breaks down texts into individual sentences and words.

While vectorization still treats texts relatively simply, tokenization goes a step further. This will be presented in the following subsection.

2.1.5.3 Tokenization

After the documents to be indexed and the textual units they contain have been identified, the tokenization process follows [22]. It is first important to clarify that the term token refers to a "distinguishable unit" of a text. More precisely, a token is a set of characters that have been meaningfully grouped into a semantic unit. In the simplest case, a token corresponds to a word. However, the more general term token also includes character strings such as "127.0.0.0" or "C++". In contrast, a token that has been normalized in a certain way is referred to as a term. With reference to the example in Fig. 2.7, valid terms could be, for example, "friend" or "roman". In this example, normalization would mean that all characters are lower-cased and reduced to the singular form.

The tokenization process is characterized by splitting text into its semantically coherent units (tokens) and, additionally, by removing unnecessary punctuation marks. Figure 2.7 illustrates this procedure. The blue arrow describes the mapping process from the given text to the tokenized text.

In the simplest case, a text is thus split into individual words in this process, and its punctuation marks are removed. However, this seemingly trivial requirement also contains its pitfalls. For example, what is a good split for the following text fragments and terms?

- O'Neill isn't amused
- co-occurence
- Apple Macintosh
- 27.02.1976
- 127.0.0.0
- C++
- Donaudampfschifffahrtselektrizitätenhauptbetriebswerkbauunterbeam tengesellschaft
- L'ensemble

In general, for some of these examples, it holds that they are unlikely queries for an information retrieval system and can probably be omitted during indexing (although for very specific requirements, it is conceivable to search for all documents containing certain years, IP addresses, or company names). The recognition of these special tokens is commonly performed using natural language analysis, so-called entity recognition methods. For the other examples in this list, the

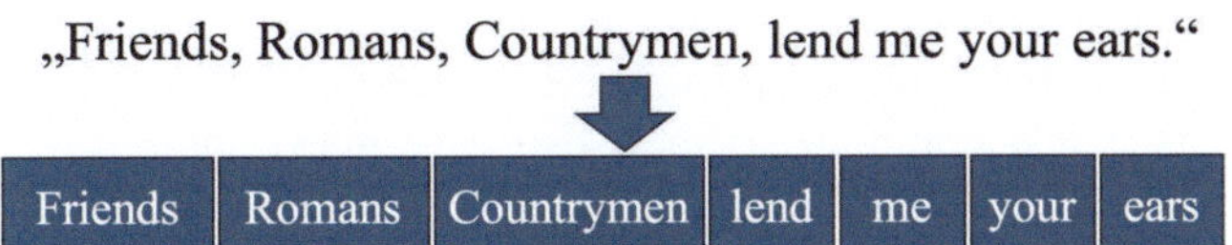

Fig. 2.7 Tokenized text

specific interpretation and processing depend on the rules of a particular language. Examples include:

- **Hyphens** are generally used in English to connect two or more words that precede a noun. For example, "some state-of-the-art articles in IR". Intuitively, this example would be split into individual words, whereas example 2, "co-occurence", should rather be indexed as a single word. In French, the use of hyphens is even more widespread and is generally used to indicate a connection between words (e.g., "couvrelit", "quatre-vingts", or "non-fumer"). The different handling between languages makes it urgently necessary to correctly identify the language used and take it into account during processing. Since the use of hyphens cannot be unambiguously used to split words, a decision must be made individually, using available rules or methods. For example, entity recognition methods can be used to identify city names such as Los-Angeles or New-York and index them together on this basis. Or rules can be applied that allow short compounds and treat them as a single word during indexing (as is the case with "co-occurence").

- An **apostrophe** is used in French to abbreviate articles when the following word begins with a vowel ("l'ensemble" vs. "le ensemble"), or it is used to replace the last vowel of a word when the following word also starts with a vowel (e.g., "Ce sont mes amis" becomes "C'est mon ami" or "L'âge"). In English, on the other hand, an apostrophe is often used to contract a noun or pronoun with a verb (e.g., "It's a dark night" or "You're not supposed to be here"). It can also be used to indicate possession (e.g., "a ship's captain" or "Mike's coat").

- **Compounds** can, among other things, lead to long word creations in German. Among the examples above is one of the longest compound German words. Words of this length are unlikely queries, but their components carry relevant information for search. To index such compounds meaningfully, they should therefore be broken down into their components. For this requirement, machine learning approaches or heuristics can be used.

In addition to the different characteristics of European languages, there are also similarities between various languages. One example is the occurrence of tokens that are used disproportionately often. These include, among others, articles. Frequently occurring tokens contribute little to a unique representation of a document due to their excessive use. Their discriminative power is therefore poor. One solution is to mark such tokens using so-called stop word lists and exclude them from the indexing process. In times of limited computing and storage capacity, this saving led to significant performance improvements. Nowadays, the use of stop word lists is less common, as they also have disadvantages. Disadvantages include, for example, phrase-based queries such as "King of Scotland", verses/song titles like "let it be" or "to be or not to be", or queries based on a relation, such as "flights to London". Modern compression methods compensate for the performance loss when using stop word lists [22].

Tokenization is thus used to split a text into individual fields (e.g., words) using linguistic and syntactic properties.

Tokenization uses properties of language and syntax to split texts into fields.

2.1.5.4 Linguistic Analysis

In addition to the challenges already introduced regarding tokenization, another problem is the normalization of tokens. A normalized token will be referred to as a term in the further course of this script. The motivation for standardization includes, among other things, the inconsistent use of abbreviations, such as U.S.A and USA, but also language features such as the use of special characters. This applies, for example, to the use of accents in French (e.g., "théâtre", "préfèrent", or "sœur") or the use of umlauts in German (e.g., Brüder, Universität, or Öl). In a normalization process, all these language features would need to be converted into a uniform form. For the use of umlauts in German, for example, the rule can be applied to encode all umlauts with the corresponding vowel plus an "e" (e.g., Mädchen becomes Maedchen and Öl becomes Oel). The capitalization of words also leads to inconsistent word representations in different languages. When searching for "Autos", one would certainly expect the same results as when searching for "autos". A commonly used solution is to simply map all letters of a word to lowercase [19].

Another fundamental point in preparing text for indexing is the normalization of inflected word forms (inflected words). Examples include conjugations, e.g., in German "geht", "gehe", "gehen" or in English "go", "went", and "gone", or plural forms such as "Autos" and "Auto" or "cars" and "car". To standardize these different forms of words, a reduction to the base forms is implemented. For this purpose, so-called stemming and lemmatization methods are used. While stemming methods are used to reduce inflected words to their word stem, lemmatization methods are used to map inflected words to their dictionary equivalent. To gain a deeper understanding of how these methods work, some of these methods will now be presented as examples:

The **n-gram method** is a fairly simple but computationally intensive stemming method. It is based on calculating a similarity measure between two tokens and is therefore language-independent. The assumption is that the probability that two tokens belong to the same group increases the more similar the two tokens are. Tokens can thus be assigned based on their group membership. The calculation of similarity is based on the assumption that similar tokens have a high number of common n-grams. An n-gram is a sequence of consecutive characters in a token. When applying the method, n = 2 is usually used. For the tokens "Information" and "Informative", the following decomposition into bigrams applies [23]:

- Information → `in nf fo or rm ma at ti io on`
- Informative → `in nf fo or rm ma at ti iv ve`

Let A be the number of unique bigrams in the first word and B the number of unique bigrams in the second word, and C the number of identical bigrams in both words, then the so-called Dice coefficient S_{Dice} provides a measure of the similarity of the two words:

$$S_{\text{Dice}} = \frac{2C}{A + B} \tag{2.1}$$

For the above example, the Dice coefficient would be $S_{Dice} = \frac{2 \cdot 8}{10 + 10} = 0.8$.

The **Porter stemmer** is a very widespread and well-known stemming algorithm. It is based on the successive recursive processing of tokens until they have a minimal number of syllables [24]. Each phase contains a set of reduction rules of the following form: `<Suffix>` $\rightarrow$ `<angepasster Suffix>`. Each phase and the rule applied within it reduces the token under consideration to a minimal representation by rule-based adjustment of the suffix. The basic idea is to examine tokens for the occurrence of certain patterns and perform a substitution. As an example, one of the rules given by Porter is introduced:

- `SSES` $\rightarrow$ `SS`
- `IES` $\rightarrow$ `I`
- `SS` $\rightarrow$ `SS`
- `S` $\rightarrow$

If this rule is applied to the token `CARESSES`, this token is reduced to `CARES`. For `PONIES`, the result would be `PONI`. These examples already show that the quality of a Porter stemmer depends on which of these rules are implemented. It also quickly becomes clear that such methods also entail a certain degree of inaccuracy. Stemming methods can generally tend toward either overstemming or understemming. Overstemming refers to the case where two different tokens are mapped to the same stem. Understemming, on the other hand, refers to the case where two words with similar stems are not mapped to the same stem.

Lemmatization refers to a method that attempts to perform analyses based on dictionaries in order to reduce tokens to their base form, e.g., `computing` $\rightarrow$ `compute` or `mice` $\rightarrow$ `mouse` [25]. As an example, the so-called StaLe method can be cited here, which is based on a rule-based approach that calculates suggestions for suitable base forms of a token. The rule base is created using a representative sample of the data set to be analyzed. The basis for creating the rules is the identification of a set of inflected tokens and their respective word stems. The rules are applied to the input string during processing. Figure 2.8 shows the basic procedure using the Finnish word "autoja" ("car" in English). In this example, the rule "oja o 100%" is found for the input. This means that the syllable "oja" can be replaced by "o" with 100% confidence.

Stemming and lemmatization methods thus ensure that the extracted terms of a document are reduced to a uniform base form as much as possible, thereby achieving an initial comparability of the terms. Both statistical methods (such as the

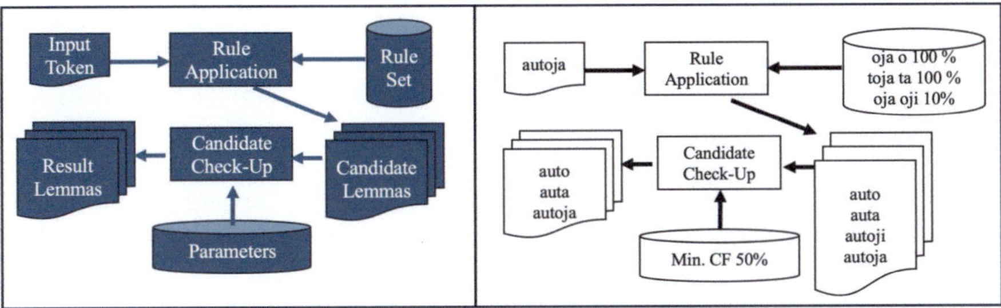

Fig. 2.8 Example of lemmatization

Table 2.5 Incidence matrix

	Anthony and Cleopatra	Julius Caesar	The Tempest	Hamlet	Othello	Macbeth
Anthony	1	1	0	0	0	1
Brutus	1	1	0	1	0	0
Caesar	1	1	0	1	1	1
Calpurna	0	1	0	0	0	0
Cleopatra	1	0	0	0	0	0
mercy	1	0	1	1	1	1
worser	1	0	1	1	1	0

n-gram method), grammatical methods (such as the Porter stemmer), and analytical methods (such as dictionary-based lemmatization) are used for this purpose. The resulting set of terms can then be used for further steps [19, 24, 25].

Linguistic analysis enables the normalization of individual tokens.

2.1.5.5 Inversion of the Index

The index of a traditional book consists of entries of words and references to the pages on which they appear. A very simple and obvious way to build the index of digital text documents in a similar manner is to determine the set of all words used and appropriately record their occurrence in the respective document. This can be implemented, for example, using a simple incidence matrix. An incidence matrix or node-edge matrix is a representation of the nodes and edges of a graph in matrix form. The elements of the matrix indicate the relationship between nodes and edges of the graph. If the entry is 1, the term is contained in a document; if the entry is 0, this connection does not exist. In this application case, the entries of the matrix would indicate whether a term x is an element of document y or not (see Table 2.5) [20].

For small data sets, this approach may be appropriate, but for large data sets, it is not suitable, since, in particular, many of the words contained are likely to occur only rarely and the matrix would have many zero entries. Matrices with many zero entries are also called sparse matrices. A solution for more efficient implementation of the index for IR applications is to use only the entries of the matrix that are actually occupied. This requirement for representing term/document membership can then be implemented using a list. Such an index is referred to in IR as an inverted index. Since a list can be designed in different ways, some general requirements for the inverted index are first described [19, 20]:

- The index should serve as a lexicon in which it is possible to look up in which documents a term appears.
- Looking up terms and accessing documents should be implemented quickly and in a memory-efficient manner.
- It should be possible to answer not only queries that evaluate the mere presence of a term in a document, but also substring matching should be possible on this basis.

An inverted index thus comprises the set of all processed tokens of a document collection. Together, these form the lexicon and are referred to here, as part of the lexicon, as terms. Each term in the lexicon has a reference to the documents in which the term occurs. The first two left columns of Table 2.6 show a naive implementation of the inverted index, corresponding to a simple mapping of terms to documents. However, access to an index prepared in this way is not really efficient. Optimization for fast access to the elements of the lexicon can be achieved by alphanumeric sorting of the lexicon entries or by grouping terms that occur multiple times. In this process, identical terms of a document are merged and then grouped. Each of these index entries that specifies the occurrence of a term in a document is called a posting, and a set of such postings is called a posting list (see Fig. 2.9).

An inverted index can be very large and thus inefficient in its original form. Therefore, a number of ways to optimize the index will be presented below. An optimal index is an important prerequisite for implementing high-quality and efficient IR processes.

Inverted indexes are the basis for search and access operations.

Table 2.6 A simple example of term distribution and term frequency

Term/Doc	Information	Retrieval	Power	Fun
DOC1	1	4	0	2
DOC2	0	3	1	1

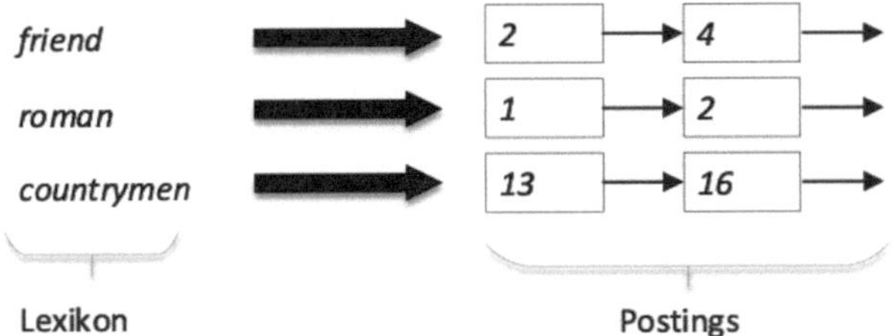

Fig. 2.9 Example of an inverted index

2.1.5.6 Weighting of Terms

As already mentioned in the linguistic analysis of text, not all tokens in a document contribute equally to the search. Stop words, for example, generally do not provide a suitable basis for determining whether a document is relevant to a query compared to another document. The reason for this is that stop words appear far too frequently in the documents of a text database and therefore do not have any discriminative power for a single document. However, this does not apply to terms that do not occur in an inflated manner. Another piece of information that can therefore be added to the inverted index is a measure of the discriminative property of a term in the form of a weighting. In other words, the basic idea of weighting methods is to assess how much influence a term should have in calculating the relevance of a document [8, 26].

The **term frequency** (TF) is an indicator of how often a term appears in a document. It refers to the number of times a term is repeated in a text. The following assumption is made: the more frequently a term appears in a text, the better the term describes the content of the document. In the simplest case, *TF* is thus an element of the natural numbers, including zero. In this case, it is simply counted how often a term appears in a document. A variant of the simple *TF* calculation is to extend the calculation by normalization. Normalization is achieved by dividing the *TF* by the maximum frequency (*freq*$_{\mathrm{max}}$) of the most frequently occurring document in the data collection:

$$TF_{\mathrm{norm}} = \frac{TF}{freq_{\mathrm{max}}} \tag{2.2}$$

Table 2.6 shows a simple example in which $freq_{\mathrm{max}} = 4$ applies.

However, the term frequency *TF* has the crucial disadvantage that the calculation does not take into account whether the terms in a database are distributed more evenly or more concentrated across the contained documents. Therefore, the **document frequency** (DF) is often also considered as an additional metric. This compensates for the disadvantage of *TF* and evaluates the occurrence or distribution of terms in a document collection. The number of documents in which the term *t* appears (document frequency) is denoted as DF_t. The **inverse document frequency** (IDF) is the inverse measure of *DF*. Unlike *DF*, it assigns a high weight to terms that occur only rarely in the collection. One possible formula for calculating *IDF* is to use the total number of documents N as a normalization factor and to calculate the inverse of this ratio: $\frac{N}{DF}$. To limit the effect of very frequent occurrences, the calculation can also be logarithmized, resulting in:

$$IDF_t = \log \frac{N}{DF_t} \tag{2.3}$$

From the indicators introduced so far, a widely used application in IR, the so-called **TF/IDF method,** can be derived. The basic idea of this method is to relate the relevance of a term within a document (i.e., its discriminative property) to the relevance of this term within all documents, thus formulating a statistical calculation basis for the index that is specifically tailored to the collection. The TF/IDF corresponds to the product of *TF* and *IDF*. It then assigns a high weight to a term if it appears frequently in only a few documents of the collection. It assigns a low weight if a term appears frequently in a document or in many documents of the collection. With these properties, TF/IDF is an important tool for assessing the discriminative properties of terms in a document. Manning [27] introduces a simple statistical calculation with which the relevance of a document for a query can be calculated using TF/IDF. The TF/IDF value is often also called the *score*. The assumption here is that the index holds the *TF* and *IDF* for all terms in the collection. The relevance is then obtained as the sum of the weights of the terms that appear in both the query q and a document d, as follows:

$$Score(d,q) = \sum_{t \in q} TF \cdot IDF_{t,d} \tag{2.4}$$

Or, in extended form:

$$Score(d,q) = \sum_{t \in q} TF \cdot \log \frac{N}{DF_t} \tag{2.5}$$

The integration of the logarithm function now makes it possible to use threshold values for index generation as well (see Fig. 2.10). The term "Upper cut-off" here refers to the threshold above which terms are no longer relevant for the indexing process because they occur too frequently within the collection. Terms whose frequency falls below the "Lower cut-off" are irrelevant because they occur too rarely. The shaded area "Significant words" then contains those terms whose indexing quality within the collection is best. Here, the suggested Gaussian curve becomes apparent, indicating that the "resolving power," i.e., the efficiency of the index term, is highest within this area. Various studies show that the statistical TF/IDF method is still superior to other methods, e.g., from the field of machine learning, and is therefore used in many IR applications. Thus, the set of terms used to index a collection can now be calculated and further optimized using the threshold values [26, 28].

> TF/IDF is an important metric in Information Retrieval.

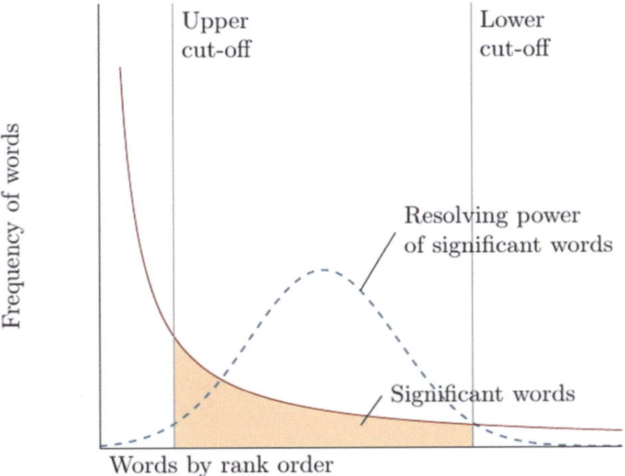

Fig. 2.10 TF/IDF threshold values [26]

2.1.5.7 Optimization of Lexicon Access

The lexicon is needed to enable the lookup of terms in an index. It provides various information, including posting identifiers of documents, pointers to associated posting lists, and statistics required for calculating term weights. In addition to optimizing a query, another well-known way to improve efficiency in query processing is to adapt the structure of the lexicon used. Although even the simple inverted index is sorted alphanumerically, the question remains how an index can be implemented to be both storage-efficient and fast to access. Prominent solutions for implementing efficient access methods are hashing techniques or tree structures.

Tables, such as the presented inverted index, can be used and extended more quickly with computer science methods if they are transferred into another data structure that allows for a simple or fast mapping to be calculated. The use of **hash functions** is a frequently used solution in computer science to achieve this. Generally speaking, a hash function is a mathematical rule that allows a unique and fixed value to be calculated for a data record (see Fig. 2.11). This mapping is represented in the form of a table, the so-called **hash table.**

The rows of the hash table are also called slots. Each slot maps a value to a hash. An optimal hash function is therefore a function that distributes all available data records to all slots without collisions [29].

To create a lexicon that can be accessed efficiently for indexing in IR, hashing methods are used to map the terms of a lexicon to a numerical value based on hash functions, thereby building a table that forms a mapping between lexicon entries (as well as their pointers to the corresponding postings) and hash values. For lookup in such a table, the hashes for the words in the query must then be calculated accordingly. Disadvantages of this method are that collisions can occur when calculating a hash (two or more values are mapped to the same slot), and

Fig. 2.11 A simple hash function

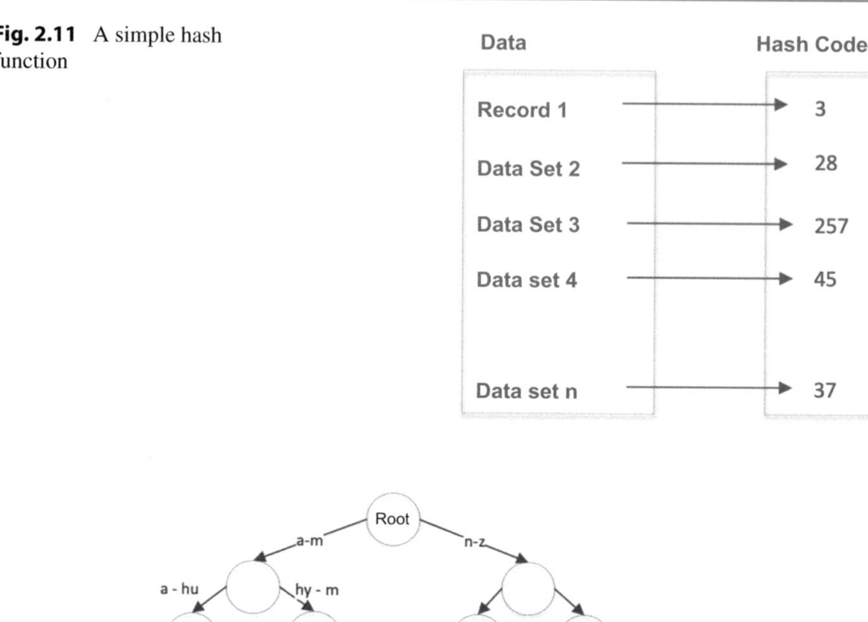

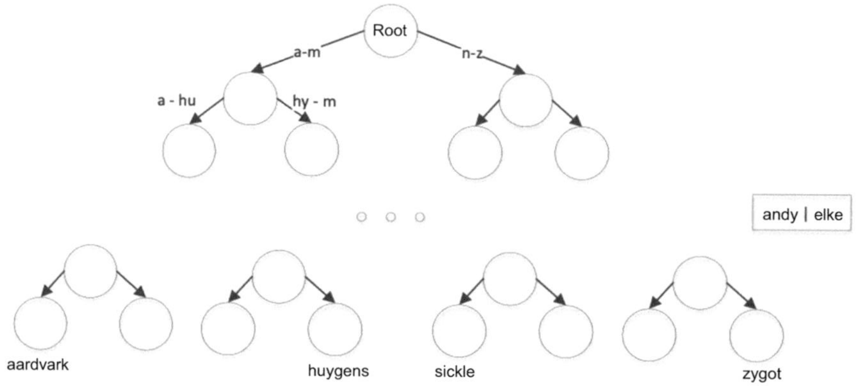

Fig. 2.12 A simple binary tree

also that very similar words can be mapped to very different hashes, making a search for similar words very costly to implement. A simple approach to avoid a collision is to assign data mapped to the same slot sequentially to the next free slot. The method must, of course, also be applied to access. This means that if a slot does not contain the expected value, a sequential search must be performed to find where the value was stored [30].

There are potentially many possible variants of a hash function that can be applied to this requirement. The challenge is to find a function that maps the existing set as collision-free as possible.

Tree structures are another option for implementing a lexicon. The simplest approach for implementing a lexicon based on a tree structure is to use a **binary tree.** The nodes of such a tree hold criteria for deciding which branch to follow to the next lower nodes or leaves (see Fig. 2.12). The problem with binary trees is that they tend to become unbalanced. Queries in an unbalanced tree, however, are costly.

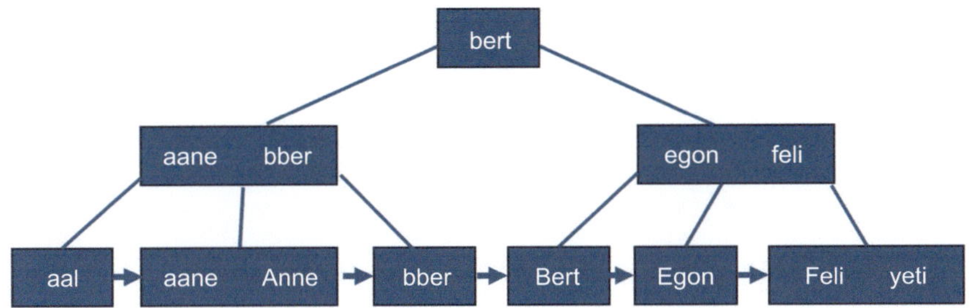

Fig. 2.13 A simple B+-tree

B-trees, on the other hand, are self-balancing trees whose nodes can have n children (potentially more than two). For implementing a lexicon, this means that the words contained in the lexicon are divided into ranges that lead via a unique branch to a leaf node, which then contains the searched word. Each node covers a certain value range, which determines which path in the tree is followed. The rule is: the right path indicates smaller values and the left path larger values. Therefore, it is also important that the underlying alphabet is ordered. Each insert/delete operation leads to a rebalancing of the tree. The challenge with this approach is to create a balanced tree. An unbalanced tree that branches very widely in some places inevitably leads to long decision paths. In addition to the simple B-tree, there are further developments that are better suited for representing a lexicon. The **B+-tree** builds on the introduced approach of the B-tree but extends its structure with a crucial rule. Every B+-tree contains its values only in interconnected leaves (see Fig. 2.13) [31].

The nodes thus serve only to decide which branch to follow during the search. An advantage of the B+-tree is, among other things, that due to the linking of the leaf nodes, it is not always necessary to traverse the entire tree. This is particularly advantageous when a specific range of the lexicon is to be examined.

Tree structures also allow the implementation of so-called **wildcard queries.** A wildcard query is, for example, the search for all terms that begin with the three letters `Mon*` (e.g., money or month). This type of wildcard query is also called a trailing wildcard query because only the last letters of the term are unknown. Possible forms would of course also be `Mo*y`, `*day`, or `M*nd*`. Trailing wildcard queries are also relatively easy to implement with B-trees. Another possibility is to use a reversed B-tree, for example. This is constructed so that each path from a leaf to the root reproduces the term in reverse order. For the term `lime`, this would be `emil`. In general, the better the later search pattern of queries is known, the better index structures can be created to efficiently support this search. Since determining the search result is, in practice, the MMIR aspect that should be implemented with the best possible performance, a great deal of effort is often invested in the calculation and organization of index structures. Typically, there is not just a single index, but multi-level, parallel, nested index structures that are already optimized for various query options. At runtime, the MMIR system then selects a processing plan based on the existing index structures that can determine the result set as quickly as possible [19, 31].

> Hash functions and tree structures optimize index access.

2.1.6 Textual Representation of Multimedia Features

It should already be noted here that not only text documents, but in general any multimedia objects can have features that can then receive textual representations. In the area of text documents, these are the terms presented here. In the area of other multimedia objects, these can be additional features, for example:

- Images: used colors, EXIF metadata (such as location, camera model, aperture setting), recognized objects (e.g., person, dog, cat, mouse), landmarks (e.g., Eiffel Tower), resolution, image format, etc.
- Audio: title, artist, label, spoken text (speech-to-text analysis), pitch, beats per minute, speaking rate, recognized sounds or keywords ("Hey Siri"), file size, codec, etc.
- Video: all information from the individual images and audio tracks, as well as metadata (e.g., MPEG-7), scene recognition, object recognition, etc.

In addition to these more general features, there are also features from specific problem domains. For example, in the medical field, specific diagnostic features are often required, as shown in Fig. 2.14 using a mammogram [6, 32].

Typically, these features are determined by feature extractors. Each multimedia feature has a natural or artificial textual representation. Natural representations arise, for example, when the text of an audio file is transcribed, an image is

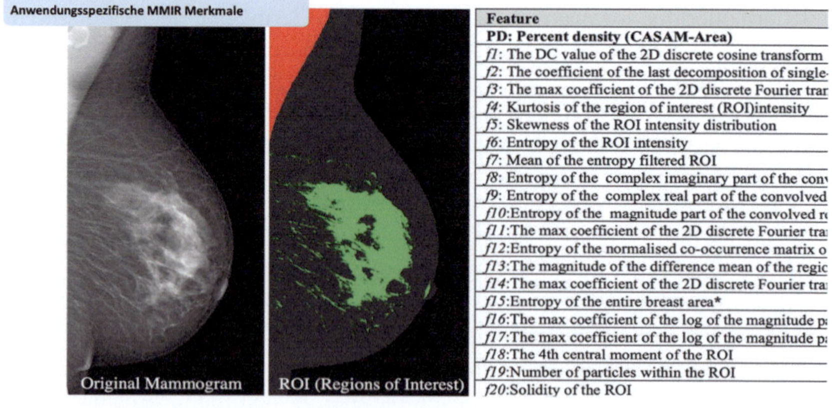

Feature
PD: Percent density (CASAM-Area)
$f1$: The DC value of the 2D discrete cosine transform
$f2$: The coefficient of the last decomposition of single-
$f3$: The max coefficient of the 2D discrete Fourier trar
$f4$: Kurtosis of the region of interest (ROI)intensity
$f5$: Skewness of the ROI intensity distribution
$f6$: Entropy of the ROI intensity
$f7$: Mean of the entropy filtered ROI
$f8$: Entropy of the complex imaginary part of the conv
$f9$: Entropy of the complex real part of the convolved
$f10$:Entropy of the magnitude part of the convolved r
$f11$:The max coefficient of the 2D discrete Fourier tra
$f12$:Entropy of the normalised co-occurrence matrix o
$f13$:The magnitude of the difference mean of the regic
$f14$:The max coefficient of the 2D discrete Fourier tra
$f15$:Entropy of the entire breast area*
$f16$:The max coefficient of the log of the magnitude p:
$f17$:The max coefficient of the log of the magnitude p:
$f18$:The 4th central moment of the ROI
$f19$:Number of particles within the ROI
$f20$:Solidity of the ROI

Fig. 2.14 Medical features of mammograms [6]

processed using text recognition, or objects in a video are named by object recognition mechanisms. Artificial textual representations arise, for example, when the recognized features are given a name or description (see Fig. 2.14). In both cases, however, every MM feature can be represented textually. Thus, all concepts introduced so far can also be applied to MM features.

If, for example, you have an image collection as a starting point, the contents of the images can be extracted as features using object recognition, represented textually, and this representation can then be statistically calculated using TF/IDF to establish similarity measures between the images. For audio collections with spoken text, text-to-speech algorithms can convert the audio files into text documents; for music collections, the metadata (title, artist, tempo, pitch, etc.) can be used to provide a textual representation of the features.

The following example shows the application of the TF/IDF algorithm using an image collection: As a starting point, a set of similar images was used, all of which have a similar structure and content and would therefore exhibit similar MM features (Fig. 2.15).

If object recognition algorithms are now applied to this collection, these contents are recognized and represented textually. The contents and their relationships can also be represented in the form of a matrix (technically, the adjacency matrix of the feature graph). Names of the features are rows and columns, the relationships between the features are values of the matrix. Applying the TF/IDF algorithm to the textual representation of the features now leads to certain features being classified as irrelevant within the collection, depending on the set thresholds. For example, it makes no sense to search for "woman" in this collection, because every

Fig. 2.15 Image collection as an example for TF/IDF application [33]

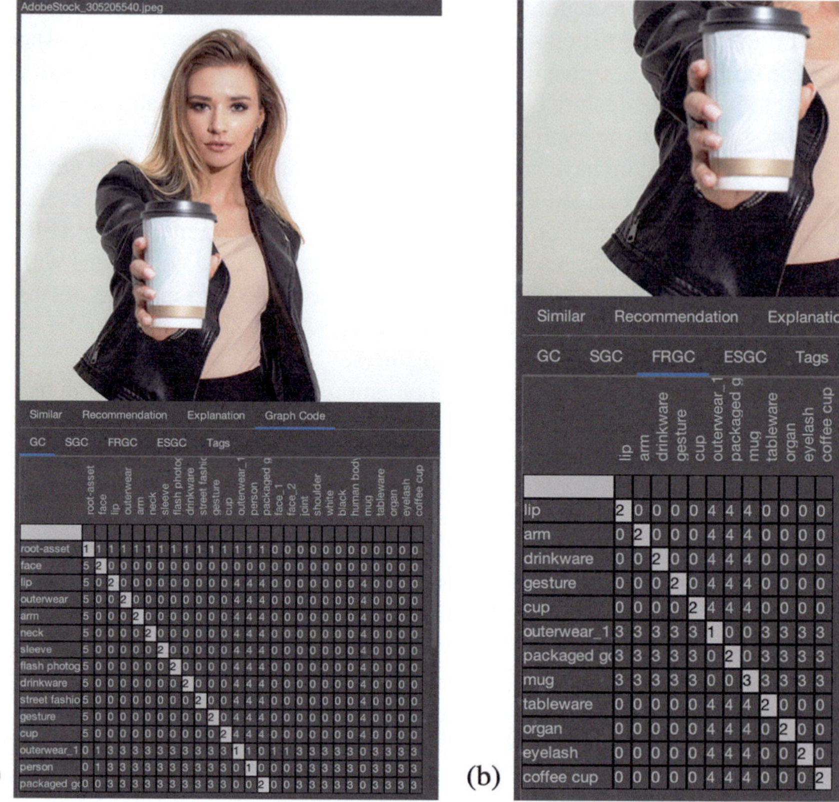

Fig. 2.16 Feature relevance through TF/IDF [35]

image in the collection shows a woman. However, it also becomes apparent that
TF/IDF is now able to identify the truly relevant features ("feature relevance") [34].

Figure 2.16 demonstrates this using an example image in which a coffee
mug is held toward the camera. Image (a) shows all recognized terms, image (b)
those that were classified as relevant by TF/IDF. All other terms were removed
by TF/IDF and—indeed—terms such as "mug," "drinkware," "gesture," "cup" are
also intuitively those that distinguish this image from the others in the collection
and thus are the discriminative features. This example shows that the concepts
introduced here from the field of document processing can be seamlessly transferred to the field of multimedia feature processing.

The example given here already shows the close interconnection between document-based information retrieval and multimedia information retrieval. Since
every multimedia feature can be represented textually, all mechanisms for processing text documents also apply directly to all other multimedia types. Therefore,
the following subsections will examine in detail how and which features can be
extracted from the different multimedia types and what textual representations

exist for them. Section 2.2 will first present methods for analyzing and extracting image features, Sect. 2.3 will show the same for audio features, and in Sect. 2.4 mechanisms for analyzing videos will finally be presented.

> All multimedia features have textual representations. The mechanisms of text analysis can be applied to them.

2.2   Image Features

Images such as graphics and photos are usually displayed on pixel-oriented output devices such as monitors or printers. This means that the graphics are composed of individual pixels (from the English "picture" abbreviated as pix and "element"), i.e., picture elements. If the pixels are sufficiently small and close together, the eye is no longer able to distinguish the individual points from each other. For the viewer, the individual pixels then merge into continuous shapes such as lines or areas, see Fig. 2.17.

The higher the density of displayed pixels, the more likely it is that the eye will no longer perceive the points as individual dots, but rather as a continuous surface. The higher the density, the better the quality of the display appears. Modern displays, such as Retina or 8K displays, have now achieved such a high pixel density that no pixels can be discerned with the naked eye. The same applies to the presentation of images on mobile devices, in VR environments (at the latest since Apple's Vision Pro headset), and also to the printing of digital media. The number of pixels displayed per image is called the resolution; for rectangular graphics, it is calculated as the number of pixels per row times the number of pixels per column. The number of colors that each individual pixel can take on, or the number of colors displayed simultaneously in an image, is called color depth. Since the eye can only distinguish a limited number of color nuances, an image with a color resolution of 24 bits is usually perceived as "natural," i.e., without noticeable

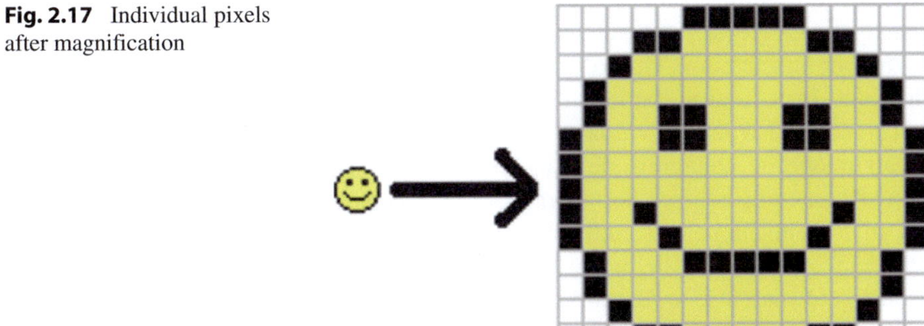

Fig. 2.17 Individual pixels after magnification

color banding. The color resolution of 24 bits is therefore also referred to as true color representation (English: "True Color"). The display of the image (English: "frame") on digital devices is controlled by the video controller or graphics chip, which continuously reads out the image memory (English: "framebuffer"). The framebuffer is a memory area in which the corresponding color value is stored for each pixel to be displayed on the monitor. The size of the framebuffer thus depends on the resolution and color depth of the image. This procedure is the same for any presentation of digital content [7, 36].

2.2.1 Image Formats

Image formats determine the way in which images are stored on a data carrier. Basically, a distinction can be made between the encoding types of pixel and vector formats. **Pixel formats** (also raster formats, English: "bitmaps" or "pixmaps") store image data in a form similar to the structure of the framebuffer. That is, the image is divided into individual pixels, and the color value is stored for each pixel. A characteristic property of pixel formats is that the level of detail in the image depends on the resolution and color depth used. The file size of an image in a pixel format is therefore directly dependent on the desired quality, which has both advantages and disadvantages: On the one hand, a high-quality image requires a lot of storage space; on the other hand, the quality of the image can be adjusted as needed via resolution and color depth. At lower image resolutions, and especially when enlarging images, stair-step effects (aliasing effects) occur more frequently. Some of the best-known pixel formats include, for example, Windows Bitmap (BMP), Graphics Interchange Format (GIF), Joint Photographics Experts Group (JPEG), Portable Network Graphics (PNG), and Tagged Image File Format (TIFF). **Vector formats** do not store images as a depiction of the entire scene, but rather as a description of the structure of the individual components of the image. These can be lines, rectangles, circles, etc. Since the framebuffer provides the image to be displayed in a pixel format for the video controller, the vector image must be converted (rasterized, rendered) into a raster format for output. The image can then be rasterized optimized for the resolution of the output medium, for example to minimize stair-step effects. If a part of the image is to be enlarged, i.e., displayed at a larger scale, that part is rasterized again at the maximum resolution of the output medium. This results in hardly any loss of quality when enlarging. However, rasterization requires computing time. Due to the modular structure of a vector image from individual components, it is also relatively easy to modify a vector image by changing, deleting, or adding descriptions. Furthermore, vector formats generally require less storage space compared to pixel formats. However, not all content can be efficiently stored in vector formats. The more geometric and "artificial" the shapes, the better the image can be stored in a vector format. Vector formats are therefore well suited for storing technical drawings, diagrams, fonts, and symbols. Vector formats are less suitable for "natural" images such as photos with many soft transitions and low sharpness. Well-known vector formats include, for example, Postscript (PS), Encapsulated Postscript (EPS), Scalable Vector Graphics (SVG), Small Web Format (Flash, SWF), and Drawing Interchange/ Exchange Format (Autocad, DXF). In recent years, a number of standard formats have been developed for storing images in different application scenarios [37]:

- **TIF/TIFF** [38]: The tag image format **tiff** was defined as early as 1986 and is now managed by Adobe. TIFF files feature lossless compression, can include metadata, and are frequently used in the fields of Optical Character Recognition (OCR), desktop publishing, or when scanning documents.
- **PNG** [39]: Files in the Portable Network Graphic format **png** also support lossless compression and are based on color palettes (24 or 32 bit). In addition, information about transparency can be stored. PNG files are often used for publishing images.
- **GIF** [40]: The Graphics Interchange Format **gif** was introduced by Compuserve in 1987 and reduces images to a 256-color palette, which typically results in very small file sizes. GIF files are increasingly being replaced by PNG files.
- **JPG/JPEG** [41]: The Joint Photographic Expert Group **jpeg** introduced the JPEG format in 1992 as a compressed format with a variable compression rate. This results in target files of various sizes, which greatly simplifies the exchange of images.
- **RAW** [42]: This format is often used by digital cameras to produce lossless images in the original format.

Table 2.7 provides an overview of the most important formats using an example image taken with a Fuji-XT3 camera [43] (RAW format, 6000×4000 pixels, 300 dpi, 50.6 MB).

> There are different image formats for different areas of application.

The human visual system is capable of distinguishing light of different wavelengths. Each wavelength in the range visible to the human eye, between 380 nm and 780 nm, can be assigned a perceived spectral color. The eye contains three

Table 2.7   Image formats and file sizes

Format	File size	Compression	Transparency	Color depth
TIFF	72 MB	Lossless	Yes	32 bit
PNG	45.8 MB	Lossless	Yes	48 bit
GIF	15.7 MB	Lossless	Yes	8 bit
JPG (Q12)	23.1 MB	Lossy	No	24 bit
JPG (Q10)	10.6 MB	Lossy	No	24 bit
JPG (Q2)	2.5 MB	Lossy	No	24 bit

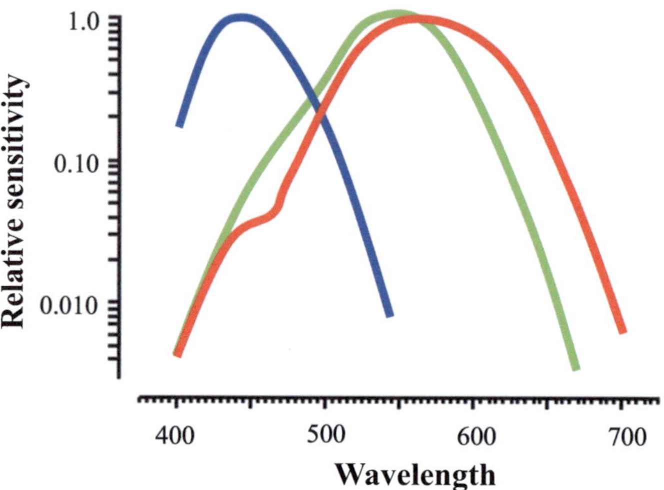

Fig. 2.18 Absorption curves for blue, green, and red cones

types of color receptors, known as cones. Each of the three types is sensitive only to a limited frequency range of light. An incoming photon triggers a specific electrical signal (receptor potential) depending on its wavelength, which is further processed in the retina and then in the brain. The frequency ranges of the three cone types overlap, so that by simultaneously considering all receptor potentials, the frequency can be inferred. It is important to note that the same color impression can be produced either by a single light frequency or by mixing several different light frequencies, as long as they trigger the same characteristic nerve stimulus. The three types of cones are called red, green, and blue cones, although this terminology is rather imprecise, since the absorption curves of the three cones are distributed very unevenly, as illustrated in Fig. 2.18 [36].

Blue cones make up only about 9% of all cones. In the very center of sharp vision, the so-called foveola, there are no blue cones at all. For practical purposes, it is especially important that with three primary colors, a large part of the visible color spectrum can be produced.

All visible colors can be represented using the colors red, green, and blue.

2.2.2 Color Models and Color Spaces

On standard computer monitors, each pixel of an image is composed of three subpixels with the three primary colors red, green, and blue (RGB). Due to the limited resolving power of the eye, a maximum of 1/60°, the three subpixels merge into a single point at a sufficiently large distance, and the three color components

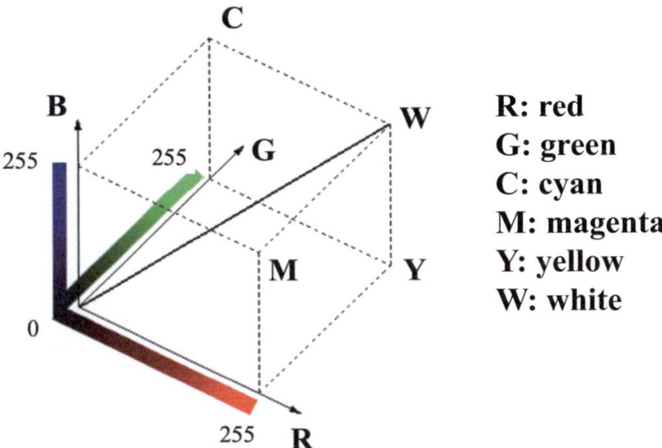

Fig. 2.19 The RGB color cube

blend into the resulting color. By varying the intensity or brightness of the components, a large part of the color spectrum visible to the human eye can be produced. Red, green, and blue together produce white. With the three parameters for red, green, and blue, a three-dimensional color space can be defined that contains all producible colors, see Fig. 2.19. A color model, in the broadest sense, determines the coordinate system of such a color space or the encoding or representation of a color [44].

In a computer's framebuffer, the color of a pixel is represented by separate values for the red, green, and blue components. For the use of the framebuffer, the RGB color model is therefore well suited, as it allows direct control of the three subpixels in the monitor via the corresponding three parameters of the RGB color model, without the need for conversion. Grassmann's laws state that any color can be represented by three arbitrary, mutually independent quantities. Thus, there are other color models, each with their own advantages and disadvantages depending on the application. The CMY(K) color model uses the complementary colors cyan, magenta, and yellow (plus black) instead of red, green, and blue, see Fig. 2.20. Complementary colors reflect all components of white light except those of their complement, or in other words, they filter out their complement.

For example, applying cyan pigments to white paper reduces the red component of the reflected light. This is therefore referred to as subtractive color mixing. Cyan, magenta, and yellow together theoretically produce black. However, since a deep black cannot be achieved in practice, pure black (K for Key or blacK) is added as a fourth color. The CMY(K) color model is mainly used in printing. Again, no color space transformation is required, as the color model matches the actual conditions. The YCbCr color model has its origins in television technology during the transition from black-and-white to color TV sets and is closely related to the YUV format used in analog television. Instead of three color parameters

Fig. 2.20 Primary colors and
their mixtures

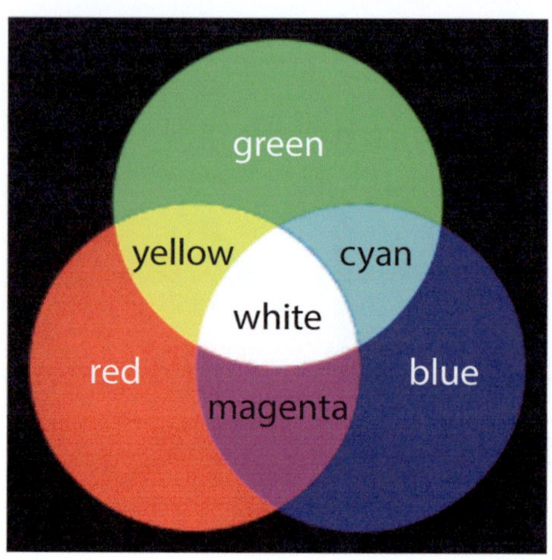

Fig. 2.21 Anti-aliasing using semi-transparent pixels

as in the RGB color model, it has two color parameters plus a brightness parameter, see Fig. 2.21. It composes a color from the components base brightness Y,
the deviation from gray towards blue or yellow Cb, and the deviation from gray
towards red or turquoise Cr. In some image formats, the YCbCr model is used to
exploit weaknesses of the human eye and save storage space by processing color

information less precisely. Perception-oriented color models attempt to accommodate the human way of perceiving color by representing colors not as mixtures of other primary colors, but by treating color information and brightness information separately. The HSV color model defines a color by its hue (H), saturation (S), and value (V). The HSV color model is often used when a specific color is to be reproduced manually. In this case, the hue is first selected, which is determined by just one parameter. In the second and third steps, the brightness and saturation of this hue are then chosen [36, 44].

An alpha channel can be thought of as an additional dimension of the color space. The alpha value determines how transparent a pixel should be for colors "beneath" it. An alpha channel makes sense when different images are layered on top of each other, known as alpha blending. If the foreground pixel has a certain degree of transparency, the background color is mixed with the foreground pixel's color by that degree. One application is so-called anti-aliasing (see Fig. 2.21). Since every output device naturally has only limited resolution, and thus a body represented by pixels can only ever be an approximation, attempts are made to smooth the pixelated edges by rendering some of the edge pixels transparent, thereby reducing their perceived intensity. This creates the impression of a more uniform structure in the eye, and the individual pixels are perceived less strongly. The fact that common image formats prescribe a rectangular base shape makes it difficult to combine arbitrary objects with arbitrary backgrounds. Here, too, the alpha channel helps by marking all areas that do not belong to the actual object as 100% transparent. In these areas, only the background is visible. Figure 2.22 shows an image with a transparent background opened in Adobe Photoshop. In addition to classic image editing, transparency is also used in web design, application design, and in the design of all types of user interfaces [36].

Fig. 2.22 Image with alpha channel in Photoshop

> The alpha channel represents a transparency layer.

Images in pixel formats require a lot of storage space: For an image with a resolution of 1024×768 pixels at 24-bit color depth, the file size is $1024 \times 768 \times 24$ bits $= 18,874,368$ bits $= 2.36$ megabytes. Since both storage space and bandwidth are scarce resources, efforts are made to keep the required storage space as small as possible. Compression, or encoding, refers to the reduction of data volume by using a different representation of the data. A distinction is made between lossless and lossy compression.

With lossy compression (irrelevance reduction), a certain loss of information is accepted in order to achieve even higher compression than with lossless compression. The original information cannot be restored. Lossy compression must always be tailored to the content and type of data. Lossy compression is mainly used for image and audio files. Here, advantage is taken of the fact that human senses have limited capabilities and certain information cannot be processed at all. For example, the limited ability to distinguish between small and closely spaced dots can be used to create the illusion of continuous figures on a monitor or paper. The ability to distinguish similar shades of color is also limited [45].

With lossless compression (redundancy compression, entropy increase), the data can be restored 100% by decompression. The amount of information remains the same. Lossless compression can be performed without knowledge of the content and meaning of the data and is therefore applicable to all types of data. Lossless compression methods include run-length encoding, the Lempel-Ziv family, Huffman coding, and the discrete cosine transform [45, 46].

- Run-length encoding (RLE) is a very simple compression algorithm. It combines consecutive identical data values and replaces them with the number of occurrences (run length) and the value itself. Run-length encoding is therefore particularly suitable for images with few colors and many solid areas. Run-length encoding is used in Windows Bitmap (RLE/BMP) [47].
- The Lempel-Ziv family, which includes LZ77, LZ78, and LZW (Lempel-Ziv-Welch), are so-called string substitution methods. They use a "dictionary" to index recurring whole words in the data set, so that only an index is stored instead of the word itself. LZW is used, for example, in GIF (optionally: TIFF and JPEG). For LZW (but not LZ77), the company Unisys, together with GIF inventor Compuserve, asserted patent claims and demanded license fees from manufacturers of software products that could write GIF files until the patent expired in 2003. The unpatented LZ77 is used by PNG [48].
- Huffman coding is an entropy encoding method. It encodes characters with a variable number of bits. A prefix-free encoding is used, meaning that no code word may be the prefix of another code word. In Huffman coding, the more frequently occurring characters receive shorter codes, and the less frequent characters receive longer codes. Huffman coders are often combined with other compression methods, such as those from the LZ family [49].

The following subsection presents the previously mentioned Huffman coding in detail, as it is used in numerous other methods.

2.2.3 Huffman Coding

Huffman coding [49, 50] is one of the entropy coding methods and is based on the idea that the required storage space for a given number of characters in binary code can be significantly reduced if frequently occurring characters require fewer digits for their representation and storage as binary numbers. To achieve this, a binary tree is constructed in which the longest paths lead to the least frequently occurring nodes. At each node traversed by the path, either a 1 or 0 is added to the binary number—depending on the direction in which the path continues. A simple example will illustrate this below.

The German sentence "Barbara mag Rhabarber" requires 168 bits of storage space when using 8-bit ASCII encoding. To reduce this space using Huffman coding, the number of occurrences for each character is first counted and recorded as a weight (see Table 2.8).

Now, each character is represented as a node together with its weight, as shown in Fig. 2.23.

Next, the nodes with the lowest weights are paired and combined into a parent node, which receives as its own weight the sum of its children's weights. This process is repeated until a binary tree is formed, as shown, for example, in Fig. 2.24.

The new binary codes for the characters are now determined by tracing the path from the root—here 21—down to the respective letter. The corresponding binary digits are noted along the way. For example, since two left edges lead to "a", the new code for "a" is "00". The same procedure is applied to the remaining characters, resulting in the following code table (see Table 2.9):

The resulting code is: "1111000100110001000100101000101110011001101000110000100111110010" and only 63 bits are now needed to store the information.

The characters encoded with Huffman coding thus have a variable bit length, which is adapted to the frequency of occurrence of each character. In the sentence "Barbara mag Rhabarber", at most 4 bits per character are used for the binary representation, saving valuable storage space without altering the content of the

Table 2.8 Number of occurrences of the characters in "Barbara mag Rhabarber"

B	a	r	b		m	g	R	h	e
1	6	4	3	2	1	1	1	1	1

Fig. 2.23 Node weighting in Huffman coding

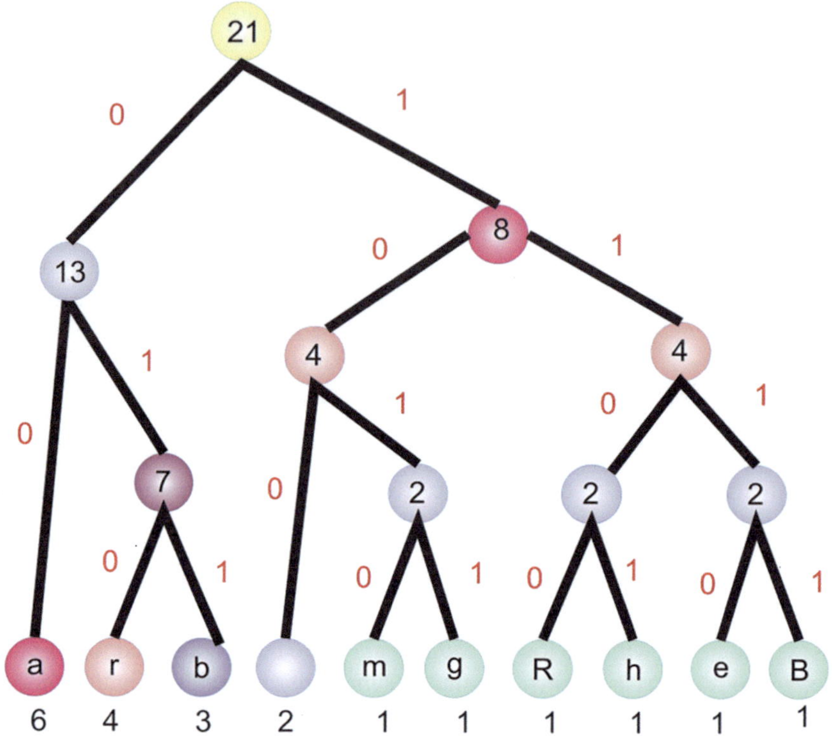

Fig. 2.24 The Huffman tree

Table 2.9 Code table for the sentence "Barbara mag Rhabarber"

a	r	b		m	g	R	h	e	B
00	010	011	100	1010	1011	1100	1101	1110	1111

message. In audio coding, the Huffman method typically achieves an average reduction of 1:2. Since the coding method is not unique, the structure of the tree must be included with the encoded file.

> Huffman coding is a standard compression method and is used in many multimedia formats.

2.2.4　Selected Image Formats

In this subsection, some of the most common image formats will now be presented in detail. This is a small selection that illustrates, on the one hand, the different methods for encoding images, and on the other hand, the possible areas of application.

2.2.4.1 Graphics Interchange Format (GIF)

GIF87a was developed in 1987 by the online provider Compuserve to enable users to download images more quickly via the online service, thanks to smaller file sizes, than was possible with the image formats available at the time, such as PCX. In 1989, Compuserve released the improved version GIF89a **gif.** GIF can store images with a maximum of 256 different colors. For this purpose, a palette is used that maps the 256 index values to one of 16.8 million possible colors, and whose colors can be sorted by relevance. A binary alpha channel allows a transparent "color" to be defined, which lets the background show through completely. By storing multiple images in a single file, it is possible to create animations or short film sequences. Furthermore, extensions can be used to control the playback behavior of the animations [40]. The interlace mode allows a rough impression of the image content to be obtained even during the loading process. GIF interlace builds up the image line by line. With each iteration step, the horizontal resolution is doubled. Figure 2.25 illustrates this concept. GIF uses the LZW algorithm for compression. Therefore, and due to the limitation to 256 colors, GIF is not suitable for photorealistic images. However, the GIF format is frequently used for drawings and symbols. In these cases, file size reductions to 50–10% of the original file are typically achieved [40].

Due to Compuserve's formerly liberal licensing policy, which allowed anyone to implement the format free of charge, GIF spread very quickly and became the standard image format on the Internet. However, Compuserve had failed to check the LZW algorithm used for potential patent claims. Unisys, which held the rights to LZW, began demanding license fees for the LZW algorithm used in GIF from 1994 onwards, together with Compuserve. This was the impetus for the development of the free GIF alternative PNG [22]. In the 2010s, GIFs experienced a surprising comeback, even though PNGs had become extremely widespread. In particular, the popularity of GIF animations on image board sites such as tumblr or 4chan led to GIF increasingly being understood as a synonym for short videos, which are used within other applications for visual illustration or as a comment function. It should be noted that these functions are often not actually provided on the basis of the "real" GIF format, and the format designation is increasingly becoming a synonym for a usage pattern (see Fig. 2.26).

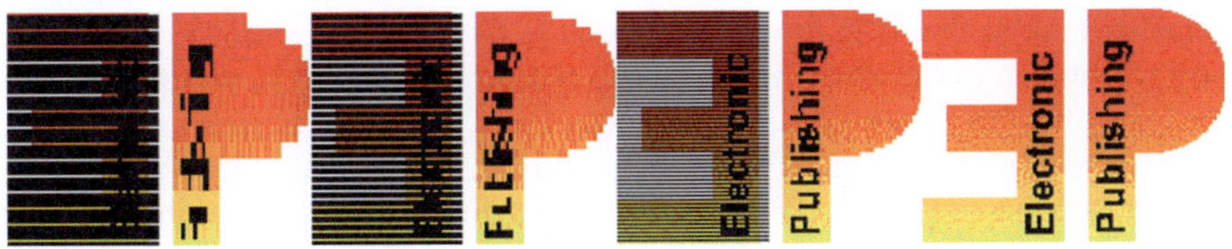

Fig. 2.25 GIF interlace iteration steps [40]

Fig. 2.26 GIF integration in
Microsoft Teams [51]

2.2.4.2 Portable Network Graphics (PNG)

After Unisys began demanding license fees at the end of 1994 for software using
the LZW algorithm, Thomas Boutell started developing an image format as an
alternative to GIF that uses only patent-free algorithms and implementations. The
Portable Network Graphics, or PNG is Not Gif (PNG) (pronounced "ping") image
format later became a recommendation of the World Wide Web Consortium and
an ISO standard. The goal of PNG was to create a replacement for the simpler
GIF as well as, to some extent, for the much more complex TIFF [39]. PNG supports all the functions that GIF offers, equally well or better. The only exception is
animation capability, which was deliberately omitted, as a separate format called
Multiple-Image Network Graphics (MNG) was intended for this purpose. PNG
supports color depths of 1, 2, 4, or 8 bits per color channel when using a color palette, true color representation with 8 or 16 bits per color channel, as well as gray-
scale representation. In addition to the RGB channels, an alpha channel with up
to 65,536 transparency levels per pixel can be stored. Furthermore, PNG offers a
progressive mode that is improved compared to GIF's interlace mode, allowing an
early impression of the image content to be obtained even with slow data transmission. For this, the image is divided into blocks of 8×8 pixels; from each block,
pixels are then displayed in the following order (see Fig. 2.27) [39].

With each iteration step, the horizontal and then the vertical resolution is alter-
nately doubled. One reason for PNG's approximately 5–10% better compression

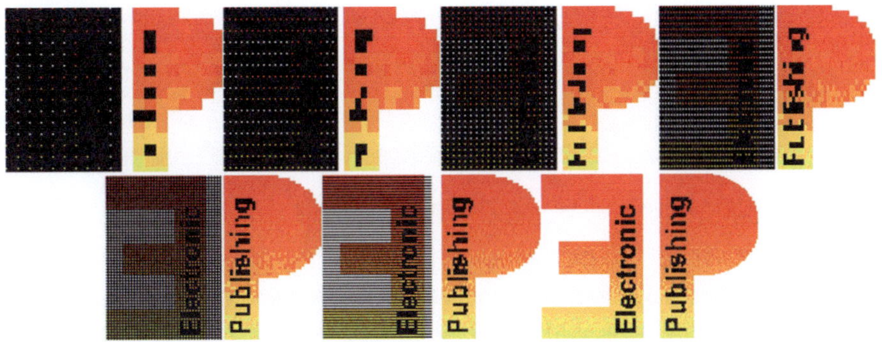

Fig. 2.27 PNG progressive iteration steps

compared to GIF is the use of filter functions. In simple terms, these replace the absolute color values of the pixels with the relative deviation from their neighbors. Since, in smooth color transitions (so-called gradients), the color values of neighboring pixels change only slightly and evenly, their differences are relatively small and similar values, which can be compressed more efficiently. For compression, PNG uses the ZLIB library, which employs the lossless and widely used Deflate algorithm. This consists of the LZ77 derivative Lempel-Ziv-Storer-Szymanski algorithm and Huffman coding, and was originally developed by Phil Katz for the free compression program PKZIP. PNG supports embedded metadata for information about the image content, as well as data for gamma correction to compensate for differences in display and interpretation between different computer systems. PNGs are streamable, i.e., random file access is not required, and they offer extensibility, especially for compression methods.

2.2.4.3 Joint Photographics Experts Group (JPEG)

The abbreviation JPEG stands for Joint Photographics Experts Group, a committee from the International Organization for Standardization (ISO) and the International Telecommunication Union (ITU-T). In 1992, the JPEG committee developed a standard for various image compression formats, officially named ISO/IEC IS 10918-1 or ITU-T Recommendation T.81 [29]. However, this standard is better known as JPEG. JPEG itself does not specify a file format as such, only the compression methods used. This missing component was later provided by the Independent JPEG Group [30] with the JPEG File Interchange Format (JFIF). The JPEG compression itself is also used by other graphics formats such as TIFF. The JPEG standard contains many different subformats, but many of these are not freely available due to licensing restrictions. The most important are Progressive, Lossless, and Sequential. Progressive defines a format that, similar to the interlace or progressive formats in GIF and PNG, builds up an image so that an overview of the image content can be obtained with only a fraction of the data. Lossless stores image data without information loss, but the compression rate is rather poor

Fig. 2.28 Image and representation of a JPEG in YCbCr format

compared to other image formats. Sequential is the best-known and by far the most common format within the JPEG standard. Almost every JPEG file uses Sequential JPEG with Huffman coding and 8 bits color depth per color channel [49]. Sequential JPEG is a standard for compressing natural images with smooth transitions and low image sharpness. Here, JPEG achieves excellent compression results thanks to its lossy compression technique. For photorealistic images, a compression rate of 90% can be achieved. For images with high contrast and sharp, defined shapes and few colors, such as drawings, diagrams, or text, JPEG is less suitable. JPEG images are limited to a resolution of 65535×65535 pixels [46]. JPEG format compression is divided into several stages: conversion from RGB to YCbCr and subsampling, application of a discrete cosine transform, quantization, aggregation, and Huffman coding [46, 49]. These stages are briefly explained below.

JPEG compression takes advantage, among other things, of the fact that the human eye resolves color less sharply than changes in brightness. To process color and brightness separately, an RGB signal is first converted into the YCbCr format (see Fig. 2.28). Since the luminance information Y has the greatest influence on the final image impression for the human eye, this can be exploited in JPEG compression to reduce image data. For this purpose, the color components Cb and Cr are sampled at a lower frequency (subsampling). This is part of the irrelevance reduction of the JPEG format **jpeg.** In JPEG compression, each of the three components of the YCbCr format is divided into blocks of 8×8 pixels each. Edge areas that do not make up the full 8 pixels are padded to 8 pixels. The two-dimensional discrete cosine transform (DCT) **dct** is then performed on these blocks. The DCT, similar to the Fourier transform, is a frequency transformation. It converts a function from the time or spatial domain into a frequency domain. The result is then a sequence of coefficients:

$$d_{mn} = \frac{1}{4} c_m c_n \sum_{i=0}^{7} \sum_{i=0}^{7} a_{ij} \cos \frac{(2i+1)m\pi}{16} \cos \frac{(2j+1)n\pi}{16} \qquad (2.6)$$

In simple terms, this means that for each field of 8×8 pixels, a series of frequency values is determined, which overall require significantly less storage space than the original RGB or YCbCr representation. In addition, this places the data in a different mathematical space, where certain calculations can be performed much more efficiently.

This transformed representation is expected to offer advantages in reducing redundancy in image information. Here, it is exploited that the color values of neighboring pixels often alternate and repeat. The recurring pixels usually differ only slightly from each other. This pattern is particularly common in textures, which JPEG compresses especially well. If one imagines the 8 brightness values of the 8 pixels of a so-called block of an image as the amplitude of a signal on a time axis, then the DCT, similar to the Fourier transform, converts a representation of amplitude/time into a representation of amplitude/frequency. This representation takes the form of 8 DCT coefficients. What is interesting here is that the higher, further right coefficients have less influence on the basic course of the signal representation than those further to the left, since they represent the higher frequencies and the human eye perceives high-frequency components less well. The result of the two-dimensional DCT on the 8×8-pixel block is an 8×8 matrix consisting of the calculated DCT coefficients. Characteristically, the values in the upper left corner of the matrix are the largest, and those in the lower right corner are the smallest [46] (see Fig. 2.29).

Because the coefficients further to the right and lower in the DCT matrix have less influence on the basic appearance of the pixel pattern, a relatively simple, individually lossy compression can be performed here by omitting the coefficients to a certain extent. This process is called quantization and involves dividing the coefficients by a constant divisor or a predefined table of values and then rounding. In other words, the value range is artificially restricted to a few allowed

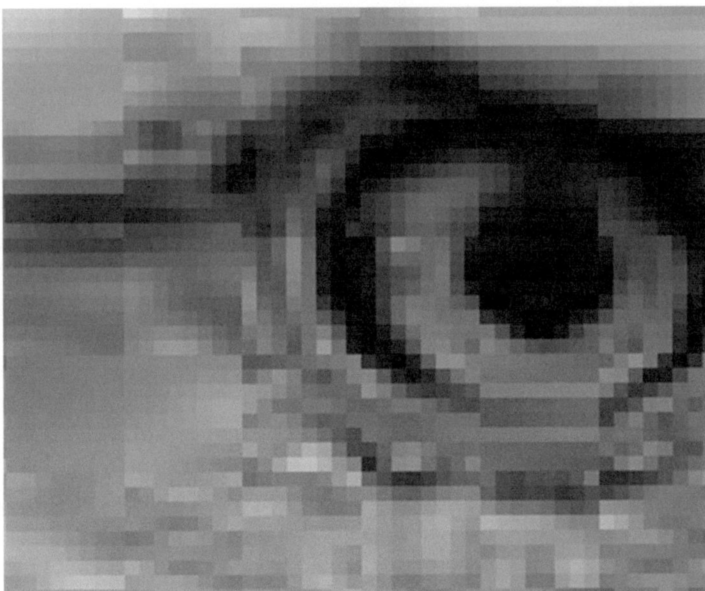

Fig. 2.29 8×8-pixel blocks in JPEG compression

Fig. 2.30 8 × 8-pixel blocks in JPEG compression as a matrix

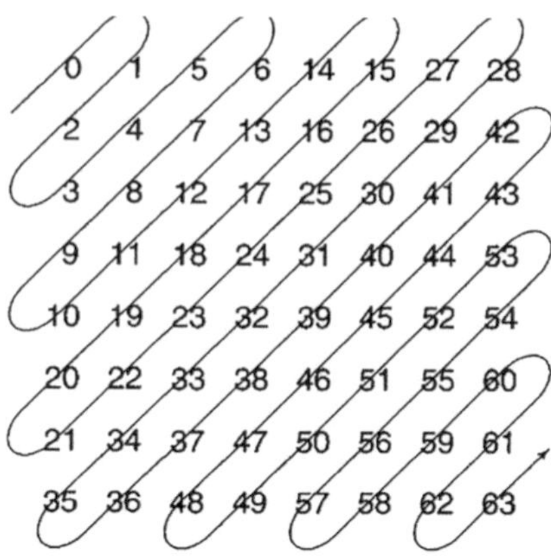

numerical values. This is the actual and variable part of the irrelevance reduction in the JPEG algorithm. The result is a matrix that only contains nonzero values in the upper left area. To make it as easy as possible to aggregate the sequence of zeros, the coefficients are grouped in a zigzag pattern starting from the top left, so that the zeros are reached at the end of the path (see Fig. 2.30). The remaining coefficients are then stored using Huffman coding [49].

In 1997, the Joint Photographics Experts Group began developing a successor to the JPEG-1 standard. Officially ISO/IEC 15444-1 or ITU-T Recommendation T.800 [29]. The goal was to create a format suitable for a wide variety of image types and application areas. JPEG-2000 attempts to eliminate the previous weaknesses of JPEG-1, further develop and improve existing functions, and add new useful features. JPEG-2000 represents advanced compression techniques optimized for efficiency, scalability, and "universality" [46]. The JPEG-2000 specification consists of a total of 12 parts, with the basic coding system described in Part 1. An attempt was made to keep all applied methods freely implementable to increase acceptance. Although not all algorithms are patent-free, all known patent holders have given their consent for free use. In the development of JPEG-2000, the wavelet method prevailed over an improvement of the DCT. JPEG-2000 consumes significantly more resources for compression and decompression than JPEG-1. Today, JPEG-2000 is mainly used in specialized fields such as medical technology, digital cinema, or as an internal graphics format for other multimedia types (e.g., PDF from version 1.5 onwards). In the consumer sector, the format has not yet caught on, as the limitations of JPEG compared to JPEG-2000 have no practical relevance for users. These include, for example, that JPEG-2000 images can be larger than 64.000 × 64.000 pixels, have up to 256 color channels, contain arbitrary metadata in XML, and can also support alpha channels.

2.2.4.4 Drawing Exchange Format (DXF)

DXF is a vector format for two- and three-dimensional graphics, created by Autodesk for its CAD software AutoCAD. In contrast to the DWG format, which also originates from Autodesk, DXF was specifically created for the exchange of vector data between different CAD programs. Autodesk has published the documentation for the DXF format, and the format may be implemented free of charge. The image data is stored as text files in 7-bit ASCII format in a file with the .dxf file extension. This makes the DXF format human-readable and facilitates the exchange of files between different computer systems. AutoCAD itself also supports a DXF binary mode, which is 25% more space-efficient and can be processed five times faster [36]. DXF supports a maximum color depth of 8 bits. A DXF file is divided into sections such as Header, Classes, Entities, etc. These are further divided into entries, which in turn can consist of group codes and data elements such as graphic elements, the so-called entities. Entities include, among others, Circle, Point, or Polyline, and represent the actual graphic elements. DXF has established itself as a de facto standard and often serves as a common denominator for different CAD programs. The first version of DXF was created in 1982 with the release of AutoCAD 1.0. Since then, newer DXF versions have always been assigned to the newly released AutoCAD versions by a designation at the beginning of the DXF file. For example, AutoCAD Release 2013–2017 is defined by the designation AC1027, Release 2018–2023 by AC1032.

2.2.4.5 Scalable Vector Graphics Format (SVG)

Another very widespread format for describing vector graphics is the SVG format, which was first published in 2001. Especially in responsive web design, it enables scalable content without loss of quality—combined with the other advantages of vector graphics over pixel-based graphics. SVG supports animations using the Synchronized Multimedia Integration Language (SMIL). One of the greatest advantages of SVG is that it is an XML-based file format and can therefore be easily created, transformed, and further processed. The basic structure of the SVG XML file is shown in Listing 2.2 [37].

```
1  <?xml version="1.0" encoding="UTF-8"?>
2  <svg xmlns="http://www.w3.org/2000/svg"
3      xmlns:xlink="http://www.w3.org/1999/xlink"
4      version="1.1" baseProfile="full"
5      width="800mm" height="600mm"
6      viewBox="-400 -300 800 600">
7      <title>Titel der Datei</title>
8      <desc>Beschreibung/Textalternative zum Inhalt.</desc>
9
10     <!--Inhalt der Datei -->
11 </svg>
```

Listing 2.2 Exemplary SVG-File.

Here, one can already see the good readability for both humans and machines, as well as the possibility of integrating metadata. SVG is based on a coordinate system (X-axis to the right, Y-axis downward), which uses floating-point numbers to describe positions and thus supports arbitrarily precise scaling. In SVG, various elements can be described, for example:

Fig. 2.31 Rendering of the SVG file from Listing 2.3

- Elements with graphical presentation (e.g., lines, circles, etc.)
- Text in definable fonts
- Grouping elements (recursive)
- Definition of colors, gradients, patterns, masks, filters (e.g., blur)
- Elements for animation
- Elements for including other language structures (e.g., scripts, style sheets)
- Embedding external documents (vector graphics or raster graphics)
- Elements for defining fonts or displaying glyphs
- Definition of transformations of elements or groups of elements

A small example illustrates the visual implementation (see Fig. 2.31) and the corresponding source code (see Listing 2.3).

```
1  <?xml version="1.0" encoding="UTF-8"?>
2  <svg xmlns="http://www.w3.org/2000/svg"
3      xmlns:xlink="http://www.w3.org/1999/xlink">
4      <path id="tp1"
5          d="M50,100 Q100,50 200,150 T350,100"
6          stroke="black" fill="none"/>
7      <text fill="black" font-size="30">
8          <textPath xlink:href="#tp1">
9              <tspan dy="3.5">Text an Pfad ausgerichtet</tspan>
10         </textPath>
11     </text>
12 </svg>
```

Listing 2.3 A Textpath as SWV File

Many programming libraries, especially in the web environment, use SVG and SMIL to create interactive websites, animations, effects, and layouts in "responsive design", i.e., size-dependent and automatic scaling.

Over time, a wide variety of image formats have been developed. Although newer formats such as JPEG-2000 and PNG generally enable better compression than their predecessors due to higher computational effort and complexity, the improvements are no longer as significant as in previous generational changes. Progress is more evident in the support of additional image types and application areas, in easier use, in eliminating limitations regarding resolution or number of colors, and in many new or improved features such as flexible data rates, error tolerance and authenticity checks, streaming capabilities, copy protection, and color profiles. However, it is questionable whether the new formats will be able to prevail over the established ones, as support must be available on all sides. Vector formats such as DXF or SVG offer significant advantages over pixel-based formats for certain applications and are also supported by a wide range of systems. Of course, the methods presented here also form an important basis for the representation of video, which, as is well known, consists of individual images. However, before the details of video formats and encoding are presented (see Subsection 2.4), another multimedia type, namely audio, will be introduced in the following subsection. After all, video, in addition to the image track, is also known to consist of audio tracks.

2.3 Audio Features

Humans are constantly surrounded by sound waves and perceive them more or less consciously. In contrast to visual perception, however, humans are unable to close their ears and simply block out continuous sound exposure. The human hearing spectrum ranges from 20 Hz to 20,000 Hz, and sound waves within these frequencies are called tones. However, there are also sound waves that are not perceived through the ears but still affect the human body. An example of this is infrasound with very low frequencies, around the order of 10 Hz. When people are exposed to these frequencies, they can experience discomfort and an increased heart rate. The entertainment industry, for example, takes advantage of this effect. For this reason, films with infrasound soundtracks are perceived as very intense [52].

Sound is therefore pressure waves in the surrounding atmosphere that propagate to varying degrees. For example, with the help of records, this sound can be permanently preserved using analog technology and, in principle, played back at any time. However, consistent playback speed is just as important as uninterrupted playback—otherwise, the sequence of sounds is quickly perceived as unpleasant. Because of these properties, audio media are time-continuous or time-critical media. The sound waves are also value-continuous, since every simple sine wave consists of infinitely many values. The art of digitization thus lies in breaking down an audio signal, which cannot be processed by a computer in its analog form, into computer-compatible digital chunks, and doing so in such a way that, when the newly created audio file is played back, at least a recognition effect occurs for the human listener.

As explained above, the sounds that reach the human ear from an acoustic environment are analog and therefore cannot be processed by a computer without further preparation, since modern information and communication technology processing methods are based precisely on digital representation methods for all types of data and information. Thus, analog sound waves must first be digitized for processing in multimedia and information systems. This process follows the three-stage model: sampling, quantization, and encoding of the sound wave, which will be explained in more detail in the following subsection.

2.3.1 Sampling, Quantization, and Encoding

During sampling and quantization, the time- and value-continuous signal of a sound wave is sampled at certain intervals (see Fig. 2.32) [53]. The sampling rate is specified as the frequency of sampling in the unit Hertz (abbreviated Hz). All values that are not captured at the time points defined by the sampling rate are discarded accordingly. After sampling, the signal is divided into a finite number of values, but these can still, in principle, have infinite precision in terms of their value range. Therefore, the second step follows: quantization.

Values that cannot be represented with sufficient precision by the computer and thus cannot be processed correctly, such as $\frac{1}{3}$ or π, are first rounded to the nearest discretely representable value. Which value is the closest is determined by the quantization intervals. What the sampling rate is for sampling, the bitrate is for quantization. Quantization introduces inaccuracies, the effect of which is also known as quantization noise. The higher the bitrate chosen for quantization, the lower the likelihood of disturbing quantization noise.

The final step of the three-stage model is encoding [50]. Here, the various quantization intervals are labeled with binary codewords. The resulting digitized signal is transmitted along with the quantization error, which, with a skillful choice of bitrate, does not fall within the audible range. Figure 2.32 provides an overview of the entire process. The method described here for analog-to-digital conversion of signals is also called waveform encoding or PCM (Pulse Code Modulation). It will be explained in more detail later on.

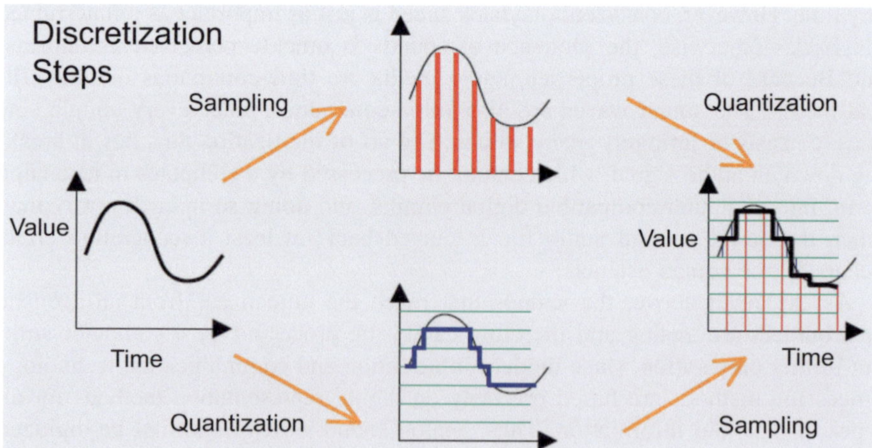

Fig. 2.32 Sampling and Quantization

There are three fundamentally different PCM methods [54]. Which method should be chosen for digitizing sound depends on the requirements for the result. Linear PCM is certainly very accurate to the original if sufficiently small quantization intervals are used, but if the file is to be transmitted over the Internet, for example, a correspondingly high bandwidth is required. In this case, all quantization intervals are of equal size. The digitized signal thus has good quality—but at the cost of a high bitrate. In dynamic PCM, the size of the quantization intervals is dynamically adjusted, for example, with smaller intervals chosen for quiet passages, resulting in less quantization noise. Dynamic adjustment can be achieved using a logarithmic function [53].

Differential PCM (abbreviated DPCM), on the other hand, requires sufficiently high computing power—but the files are smaller. The deviations between the individual sampled values are often only slight. In this method, only the difference between successive sampling values is stored. Optimal results are achieved using predictive coding, a method in which predictions about future data are made based on already analyzed data. The principle of DPCM has been further developed into adaptive DPCM (abbreviated ADPCM).

For encoding speech files, nonlinear methods are also suitable, since larger quantization intervals are sufficient for strong signals, whereas a more precise approximation is required for weak signals. Thus, depending on the application, the advantages must be weighed against the disadvantages. Determining the optimal sampling rate plays an important role in this context. Therefore, a brief mathematical excursus on the sampling theorem follows at this point [55].

> Sampling describes the rate at which the value of a signal is measured. Quantization converts the value into a discrete range.

2.3.2 Sampling Theorem

If the sampling rate is too low, the signal can only be represented inaccurately. This is countered by the desire to obtain as few values as possible so that subsequent processing does not become too extensive. The famous sampling theorem by Nyquist, Kotelnikow, Raabe, and Shannon [55] deals with the optimal sampling rate and is a fundamental theorem in communications engineering, signal processing, and information theory. The sampling theorem states that a signal band-limited to f_{max} can be exactly reconstructed from a sequence of equidistant samples if it has been sampled at a frequency greater than $2 \cdot f_{max}$.

From this theorem, it follows that a signal can be fully determined by a sampling frequency f_A that is twice as high as the highest frequency present in the signal. Let the highest frequency be f_{max}, and thus one obtains $f_A \geq 2 \cdot f_{max}$.

Since humans can hear frequencies ranging from about 20 Hz to 22 kHz, according to the sampling theorem, a sampling rate $f_{max} = 22$ kHz is fully sufficient, so a sampling rate of ≥ 44 kHz is entirely adequate for optimally sampling analog audio signals. The mathematical foundations of signal digitization are well and thoroughly summarized in [56].

2.3.3   Methods for Compressing Audio Data

After digitizing audio information, it is now available as files that can be processed and played by a computer, but historically (and even for current applications such as music streaming) these files still require a relatively large amount of storage. One minute of listening experience stored in this form already requires about 10 MB. So, for example, to store a music album with approximately one hour of playtime, about 600 MB are needed—which easily fits on a CD. As is often claimed, this is no coincidence; the size of the CD was originally designed so that Beethoven's Ninth Symphony, conducted by Karajan, could be stored on a CD-ROM.

Historically, the amount of data required for storage on hard drives, transmission over the Internet, or use in mobile playback devices was too large. Therefore, various approaches to audio compression were gradually developed and improved, which will be briefly presented in the following sections using a selection of common compression methods and techniques. It should be noted that even today, despite seemingly unlimited bandwidths, the scaling of streaming services such as Spotify, Apple Music, or Amazon Music still places very high demands on audio compression in order to efficiently and with high quality supply millions of users with digital audio signals. The Huffman coding method introduced in Subsection 2.2.3 can also be excellently used in the audio domain to compress signals [49].

The masking threshold or masking [55] refers to a method that takes advantage of a selection effect of the human ear. Anyone who has ever tried to have a conversation on a highway can relate to this effect. The voice of the person you are talking to becomes inaudible as soon as a truck passes by. Quiet sounds are thus masked by the simultaneous occurrence of loud sounds. This phenomenon is also called simultaneous masking. Sounds that cannot be perceived anyway only take up storage space and can therefore be eliminated from the file.

> The masking threshold describes that the human ear can only perceive certain combinations of tones.

The masking effect even persists beyond the actual playback time of the loud signal. An extreme example: Anyone with normal hearing who has spent a quarter of an hour next to a running jackhammer will hardly be able to hear anything for quite some time afterward. Whispered tones after such a noise disturbance do not reach the listener and can therefore be omitted. This shift in the masking threshold is called temporal masking.

Z Bark	f_u, f_o Hz	f_m Hz	Z Bark	Δf_g Hz	$10 \lg \Delta f_g^{*}$ dB
0	0				
1	100	50	0,5	100	
2	200	150	1,5	100	20
3	300	250	2,5	100	20
4	400	350	3,5	100	20
5	510	450	4,5	110	20
6	630	570	5,5	120	21
7	770	700	6,5	140	21
8	920	840	7,5	150	22
9	1,080	1.000	8,5	160	22
10	1,270	1.170	9,5	190	23
11	1,480	1.370	10,5	210	23
12	1,720	1.600	11,5	240	24
13	2,000	1.850	12,5	280	25
14	2,320	2.150	13,5	320	25
15	2,700	2.500	14,5	380	26
16	3,150	2.900	15,5	450	27
17	3,700	3.400	16,5	550	27
18	4,400	4.000	17,5	700	28
19	5,300	4.800	18,5	900	29
20	6,400	5.800	19,5	1.100	30
21	7,700	7.000	20,5	1.300	32
22	9,500	8.500	21,5	1.800	32
23	12,000	10.500	22,5	2.500	34
24	15,500	13.500	23,5	3.500	35

* 10 lg Δfg – Zunahme des Pegels (in dB) von Ausschnitten aus weissem Rauschen bei Verbreiterung des Frequenzbandes von 1 Hz auf die Breite Δfg der Frequenzgruppe.

Z = Frequenzgruppe, f_u = untere Grenzfrequenz, f_o – obere Grenzfrequenz, f_m – Mittenfrequenz, Δfg – Bandbreite

Fig. 2.33 Value table for the unit Bark [57]

The German physicist Heinrich Barkhausen (1881–1956) used the masking effect to divide the human hearing range into 24 critical frequency bands (see Fig. 2.33) [57]. He found that the width of these bands is not constant but changes with the mean band frequency. In psychoacoustics, the unit Bark is therefore used as the measure for the width of the bands.

For frequencies f below 500 Hz, one Bark equals $\frac{f}{100}$, and for frequencies above 500 Hz, one Bark is calculated as $9 + 4 \cdot \log(\frac{f}{1000})$, as can be found in [53]. Based on the signal strength in a frequency range, the duration of masking can be calculated using the Barkhausen bands, and information about masked tones can be removed without impairing the listening experience.

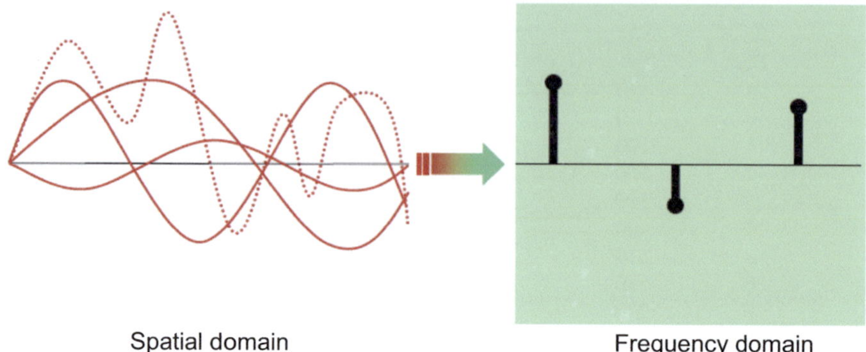

Fig. 2.34 Fourier Transform

Predictive coding is a method in which the most likely next signal is calculated from the signals already played. If the following signal does not match the prediction, only the difference from the previously assumed signal is stored. These differences require less storage space than the original signal, and thus the audio file can be effectively compressed.

Tilman Liebchen from TU Berlin used this method to develop the well-known LPAC codec ("Lossless Predictive Audio Compression"), which can losslessly compress audio files in the pac format for Unix systems. In 2004, this codec was succeeded by the MPEG-4 LAC method ("Lossless Audio Coding") and the Ogg-FLAC method ("Free Lossless Audio Coding"), which are still in use today [58].

Transform coding transfers data into a mathematically more compressible domain. A widely used method here is the Fourier transform (FT) from the time to the frequency domain [59, 60]. The sound waves are treated as polynomials, and the coefficients are analyzed for their frequency of occurrence. Smaller coefficients are omitted, resulting in a concise representation in the frequency domain (see Fig. 2.34).

The formula for the Fourier transform (FT) for $2n$ is:

$$\forall j = 0, \ldots, n - 1 :$$

$$f_j = \sum_{k=0}^{2n-1} x_k e^{-\frac{2\pi i}{2n} jk}$$

The most effective methods of this kind are the discrete cosine transform (DCT) and the fast Fourier transform (FFT), which is called "fast" because it operates according to the classic divide-and-conquer principle. Detailed information on both calculation methods can be found in [61].

In sub-band coding, the audio signal to be encoded is also transformed from the time to the frequency domain using a filter bank, which, for example in MP3, is divided into 32 frequency bands of 625 Hz each, effectively creating sub-bands of equal width.

It is assumed that only certain frequency maxima are needed for correct sound reproduction. This method is therefore also called selective frequency transformation and is particularly well suited for compressing speech files [61]. In MP3, for example, one sample is output for every 32 input samples per sub-band. The sub-bands are each further divided into 18 sub-areas using a modified discrete cosine transform (MDCT). This results in higher spectral resolution. The risk of band overlap is thus reduced, and the likelihood of aliasing artifacts is also diminished.

> There are various methods for compressing audio for different applications.

2.3.4 File Formats and Codecs

There are now a great many different file formatting methods for storing digital audio signals on appropriate storage media, and the following will discuss a selection of the most important ones. The focus here is on the method used in MP3, as the development of this method has had a particularly lasting impact on the world of consumer-oriented multimedia applications in mass markets and is therefore still the most widely used today. Due to its high relevance in the audio field, some further details on the MIDI and MP3 formats are provided below.

The acronym MIDI stands for "Musical Instrument Digital Interface" [62], which aptly describes the original idea behind this unusual audio data format. The MIDI protocol was developed in 1983 to enable communication between different synthesizers. Control signals are sent to the respective hardware, which then produces the desired tone. This means that the hardware generates the music and only the control signals need to be stored. This technique results in comparatively small files—but the quality of playback always depends on the hardware installed at the receiver.

If this method is to be used for storing audio files, another disadvantage becomes apparent: storing vocals is nearly impossible. At best, the melody can be stored. For example, the song "Daniel" by Elton John can be stored in MIDI format in just 20 KB [10], but the music lover will have to do without Elton John's voice when listening. However, this disadvantage is offset by significant advantages, one of which should be particularly emphasized here: since the tones are stored as control signals, they can be reproduced 1:1 as sheet music, and interested musicians can use programs such as Apple Logic Pro [63] to print out the sheet music for pieces stored in MIDI format (see Fig. 2.35).

Anyone who also has a MIDI keyboard or another MIDI-capable musical instrument can use it as an input device and have the notes played displayed directly as a score using programs such as Capella. To connect MIDI-capable devices, two cables with 5-pin connectors are required, directly connecting the two devices via MIDI-In and MIDI-Out. Alternatively, the computer can be connected to the keyboard via the USB interface. USB-MIDI hubs and appropriate cables are available from specialist retailers. The incoming control signals are converted by

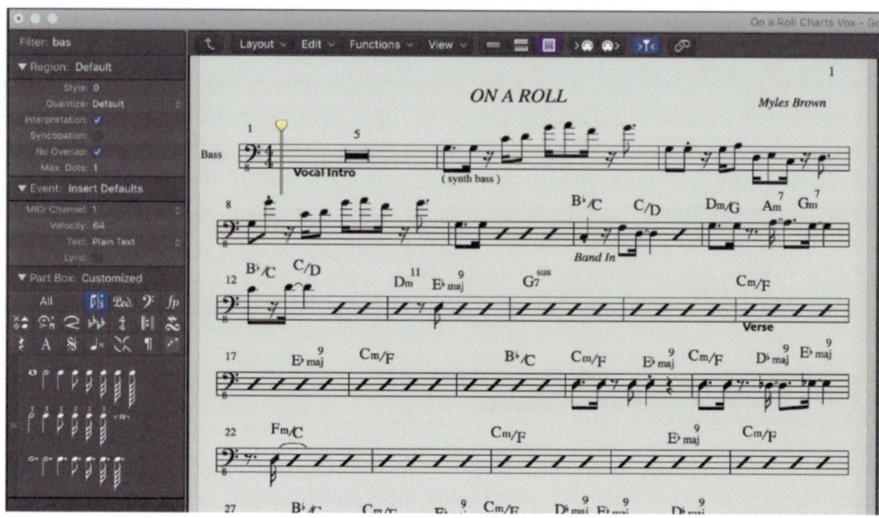

Fig. 2.35 MIDI file editing with Logic Pro [63]

the hardware into the most realistic sounds possible using wavetables. Owners of sound cards without a wavetable can supplement them by installing soundfonts and samples [63].

An extension of the standard MIDI format is General MIDI. Here, the channels are assigned to specific instruments [62]. GM1 is a standard developed by Roland in 1991 and has since been recognized and used as the lowest common denominator for standardization by various other companies. However, both Roland and Yamaha have developed additional standards—GS and XS—which led to the different devices no longer being easily interconnected. Since 1998, Yamaha and Roland have agreed on GM2 as a common standard, which is still jointly used and further developed today.

The MP3 format is currently the most widely used format worldwide for storing compressed audio files. Its full name is MPEG-1 Audio Layer 3 format or MPEG-2 Audio Layer 3 format, where MPEG stands for Motion Picture Experts Group, as this format was originally intended for compressing video data—but increasingly, only the audio codec of this format is used. The motivation for its development was to create an audio format that would allow the transmission of high-quality audio files over ISDN while saving valuable transmission time. The technical leap was intended to be as impressive as the switch from black-and-white to color television. However, when the MP3 method was developed 15 years ago, hardly anyone in Germany believed it would be possible to create a chip capable of handling such complex processes as, for example, the psychoacoustic model. It was also assumed that the costs for such a development would be immense, and so the German industry saw no chance of success for the project. Thus, while the birthplace of the first MP3 player was Erlangen—where

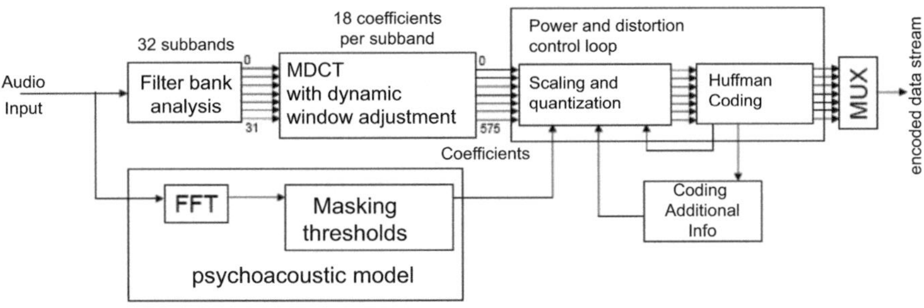

Fig. 2.36 MP3 compression scheme [55]

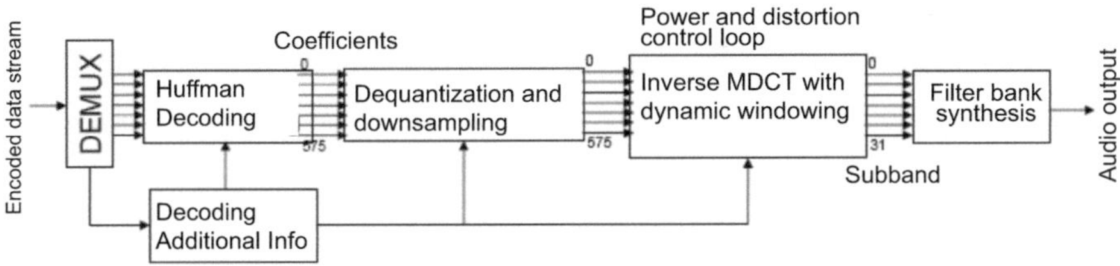

Fig. 2.37 MP3 decoding scheme [55]

Fraunhofer IIS and semiconductor manufacturer Intermetall developed the world's first model in 1994—the commercial success of MP3 initially began outside Germany.

The prototype of the first MP3 player weighed about 2 kg and stored audio data with one minute of playtime on a built-in EPROM. Although this was not yet of great practical use, it represented a breakthrough, as it was now finally possible to play music without using mechanical parts. The compression process used by MP3 is clearly illustrated by the compression scheme shown in Fig. 2.36.

As can be seen from this figure, the audio data stream undergoes a relatively complex processing sequence during encoding, but this is worthwhile in terms of the achievable compression rates, i.e., the degree of reduction in storage requirements. With MP3, enormous compression rates can sometimes be achieved thanks to variable bitrates. For example, with telephone quality at 8 kbit/s, a compression of 96:1 is achieved. For CD-quality, the audio file is reduced by a factor of 12:1 at 128 kbit/s.

Of course, the encoding must also be reversible. This process also requires a considerable amount of computation. Figure 2.37 therefore illustrates the decoding scheme for MP3 formats [57, 64].

First, the encoded data stream undergoes Huffman decoding and is dequantized. After reversing the MDCT, the undersampled spectral components, also called sub-bands, are merged. After completing the filter bank synthesis, the audio data stream is passed to the audio output and the audio file can be played.

The improved successor to the successful MP3 is considered to be MPEG2/4 AAC [62]. Another relevant and well-known codec is the Dolby codec. The company Dolby has developed proprietary audio codecs with the formats AC-1 to AC-3, which are not open. AC-1 was originally developed for HDTV; AC-2 is an improved version compared to AC-1, and AC-3 supports the multichannel sound system for surround sound with Dolby 5.1. The number 5 stands for the 3 primary channels together with the two surround channels, and the 1 for the low-frequency bass channel. This format is used in the USA for HDTV and DVD and worldwide as the standard for cinema sound, and is therefore not compatible with European HDTV, which is encoded using the MP3 method. Further information about Dolby and the various formats can be found online at www.dolby.com [62].

2.4   Video Features

Terms such as vertical refresh rate, interlaced, PAL, or composite video are unclear or unfamiliar to many. For this reason, they are introduced in a structured manner and explained in more detail in this chapter. Following this, both analog and digital video signals are examined more closely. In each case, the properties that are later relevant for video compression are highlighted. At the beginning of the last century, one of the most significant inventions of modern times was made: the cathode ray tube (CRT) was invented. With an electron beam gun, the control of the electron beam, and a phosphorescent screen, it became possible to generate moving images. Tube televisions or tube monitors always operated according to the same principle. Unlike earlier monochrome devices (e.g., black-and-white televisions), color devices consist of three electron beam guns, each responsible for one of the primary colors. The idea is that the electron beam, at a specified intensity, causes the phosphor applied to the screen to glow. To create an image, the electron beam scans the screen line by line. For a long time, a television image consisted of 625 of these horizontal lines. At the end of such a line, the electron beam must be redirected to the beginning of the next line. During this process, the electron gun is briefly switched off, as this movement would otherwise be visible. The time required for this adjustment is the horizontal retrace time. The smaller this is, the higher the horizontal refresh rate. A similar process occurs during the vertical retrace time, where the electron gun is returned to the beginning of the first line. To achieve a smooth image for the human eye, at least 25 frames per second must be generated. Through a trick, the resulting television image can be improved without increasing the number of frames, which would require higher data transmission. In this process, a single image is split and generated sequentially in two "fields": the image is interlaced. The two partial images each consist of every other line of the screen. The first field is made up of all the odd-numbered lines; the second consists of the even-numbered lines. First, the partial image starting with the first line is generated. After the electron beam has scanned all the horizontal lines of the first partial image, i.e., all odd lines, it is set to the beginning of

Fig. 2.38 Interlaced Video

the second line and illuminates all the lines of the second partial image. Together, these two half-images are called a frame. Through this interlacing process, the viewer perceives double the refresh rate; the television image appears less flickery. The conventional method of building up the television image line by line is called non-interlaced or progressive scan. In Fig. 2.38, the upper section shows the so-called "upper field," in which all odd lines are drawn first. Subsequently, all even lines of the "lower field" are drawn, resulting in a complete image for the viewer [65, 66].

A signal is analog if a change in the signal results in a proportional effect. In the case of the video signal, the luminance (brightness) changes analogously to the voltage. It follows directly from this that an analog television image does not consist of pixels, but rather the lines are a continuous variation of brightness that cannot be further subdivided. With the development of color television sets, the luminance signal alone was no longer sufficient, as it contains no information about the color of individual image elements. Color theory tells us that all colors can be mixed from the primary colors red (R), green (G), and blue (B). However, transmitting each signal individually would exceed the available bandwidth. Analog video signals can generally be distinguished by their quality as follows [66]:

- FBAS (composite video) is the lowest quality transmission method because only one signal is transmitted. Each letter stands for a specific feature: F = color signal (red, green, blue), B = image signal (light and dark tones), A = blanking signal (temporarily switching off the picture tube), S = synchronization signal (synchronization between transmitter and receiver).
- Separated video (Y/C, S-Video) separates luminance (Y, brightness) and chrominance (C, color), is a quality level higher than FBAS, and is used by S-VHS and Hi-8 recording formats.
- Component video transmits (RGB or YUV) signals.

During transmission, analog signals are affected by noise, so that in the worst case the useful signal becomes unusable. Copying analog signals also often leads to quality loss. Therefore, it makes sense to digitize analog signals. Especially in today's world, where videos are predominantly played via internet-based streaming platforms, analog signal processing usually takes place even before the actual recording. The original multimedia scene (i.e., the real world object) is by definition analog. Recording devices such as video cameras or smartphones digitize this original scene directly during recording, so the previous step of subsequent digitization is now obsolete, as cameras can already natively generate the corresponding digitized format. Nevertheless, for the evaluation of video methods, it is important to take a closer look at the topic of digitization.

2.4.1　Digitization

The digitization of video signals essentially consists of three steps: sampling, quantization, and the generation of a digital sequence of numbers/encoding. The signal is thus discretized twice: into a fixed grid of equal temporal subchapters and into a more or less fine grid of values corresponding to each point in time. The required sampling rate (sampling frequency) is determined by the highest signal frequency present. For quantization, the following applies: the more levels allowed, the higher the signal quality [67].

In addition to the previously described video signal digitized from an analog source, there is also the possibility of capturing the original video signal directly in digital form. In this case, during camera recording, the image is not converted into a continuous voltage fluctuation as with analog video, but is stored directly as a finite number of pixels. Each pixel, or more precisely, each color a pixel can take, is encoded by a binary number, which depends on the pixel depth. With higher pixel depth (i.e., more binary digits available), finer color differences can be encoded than with lower pixel depth. This is equivalent to the number of colors available for image storage. This leads to the following disadvantage: unlike analog video, a digital image loses image quality when the resolution is increased. On the other hand, digital video has the advantage that its quality is independent of the characteristics of the transmission channel, digital video can be copied without loss of quality and is practically indefinitely durable, and precise digital filters can be used, making image manipulation and effects easier and better than with analog video [66, 67].

The physical size of video files depends on several factors:

- Frame size (f in width × height) in pixels: The smaller the frame dimensions, the less storage space is required.
- Frame rate (r in fps), number of frames per unit time: The fewer frames per second, the smaller the file (15 frames per second is the lower limit).
- Color depth (c in bits): The fewer colors used, the smaller the file.
- Sound (As, size of the accompanying audio file): The lower the sound depth, the smaller the file.
- Time (t in seconds): The duration of the video.

The following formula can be used to calculate the raw video size Fs:

$$Fs = f \cdot r \cdot c \cdot t + As \tag{2.7}$$

In addition to many advantages over analog video, digital video also has significant disadvantages. Processing a digital video involves handling enormous amounts of data (e.g., when storing or transmitting digital video). With today's resolutions demanded by customers on the market, for example in the Ultra HD range (3840 × 2160 pixels), typical frame rates of 60 Hz, and a color depth of 24 bits with stereo sound (11 kHz, 8 bits), the following calculation example results for one minute of video:

$$Fs = (3840 \cdot 2160) \cdot 60\,\text{fps} \cdot 3\,\text{Bytes} \cdot 60\,\text{s} + 1,2\,\text{MB Audio} = 8.5\,\text{GB}.$$

This means that a single minute of uncompressed UHD video would no longer fit on a standard DVD (size 4.7 GB). Fortunately, since the introduction of digital SD video (resolution 720 × 576), HD video (1280 × 720), and Full HD (1920 × 1080), compression methods have been developed and standardized to ensure that the amount of data can be significantly reduced. This is, of course, existentially important for streaming platforms in particular [67, 68].

> Uncompressed video files become very large.

2.4.2   Compression

The primary goal of video compression is to eliminate any information not required in the original signal, thereby reducing bandwidth requirements without introducing perceptible differences in quality. The basis for reducing video signals lies in the presence of similarities within the image. When comparing two consecutive frames, it becomes clear that the image content changes very little from one frame to the next. Often, we can observe that these changes are solely due to the movement of individual objects. Two typical data reduction principles are entropy coding and source coding, with practical applications usually combining both. Source coding can be further subdivided into predictive and frequency-oriented compression.

Entropy coding eliminates redundancies and is typically lossless. For example, Variable Length Coding (VLC) uses Huffman coding. Probabilities are assigned to each symbol to be encoded. An algorithm then determines the encoding of a symbol using the minimal number of bits required for its given probability of occurrence. Huffman codes (see Subsection 2.2.3) have already been introduced several times in this book and also form an important basis for compression in the video domain. Further data reduction can be achieved through Run Length Coding (RLC). Many data sets consist of sequences of identical bytes. The greater their number, the better the data reduction rate. The algorithm counts repeating bytes and places a marker—one that does not occur as part of the data—before the first sequence of bytes to be repeated. **Source coding** uses irrelevance reduction, which involves trimming the encoded information. Signal components that cannot be reconstructed during later decoding are removed. Which information is considered irrelevant depends on the specific application [69].

"Prediction" or "relative coding" refers to a data reduction method based on the principle of "differential coding." Instead of storing both images in full, only one image and the difference to the next or previous image (depending on implementation) are stored [69]. **Differential Pulse Code Modulation (DPCM)** exploits large-scale regularities in an image by storing only the differences between consecutive images or between neighboring pixels within an image. For large areas, the differences between neighboring pixels are small; at edges, they are large. The process of forming differences can be described as prediction, where the next image or pixel is estimated based on previous images or pixels [70]. Using a prediction function, a prediction (estimated values of the image data) of the next image (or neighboring pixel) is calculated, but only the difference between the actual value of the next image (or neighboring pixel) and the prediction is stored. For example, if two pixels have gray values of 194 and 209, which can be encoded with 8 bits, the difference is 15, which can be represented with only 4 bits. This difference is also called the prediction error. The values estimated by the prediction function are closer to those of the neighboring pixel or subsequent image than the original values of the previous pixel or image. Therefore, smaller differences can be achieved, resulting in greater data reduction. DPCM can be applied to individual frames (intraframe DPCM) and to temporally adjacent frames (interframe DPCM). The latter increases the efficiency of DPCM, as two consecutive frames generally exhibit greater similarity than neighboring pixels [70].

Consecutive frames differ mainly in that image regions often only move rather than change. Since movements rarely change from frame to frame, the position of an object whose movement has not changed direction within a certain time interval can be predicted in the following frame with high probability. Using motion prediction from comparing the current and previous frames, motion compensation can be applied to the current frame. The frame is modified according to the motion prediction, making it much more similar to its successor, which facilitates difference formation.

In **transform coding**, data are transformed into another mathematical domain, making them more effectively compressible. One of the most effective transformations for data reduction is the Discrete Cosine Transform (DCT) (see formula 2.6), which is also used in JPEG and MPEG [41, 71]. The DCT transforms the pixel values of an image block from the spatial domain into the spatial frequency domain. Coarse image structures and slow brightness transitions (areas) are represented by low spatial frequencies, while fine image structures and rapid brightness transitions (sharp edges) are represented by high spatial frequencies. The focus is on the low-frequency components, as these occur more frequently in natural images than fine details. Omitting high-frequency components produces irrelevant errors that are often not or barely visible. For the DCT, the image is divided into square 8×8 pixel-sized blocks, and the transformation is applied to each of these blocks. The transformation of a block produces a block of the same size with DCT coefficients—analogous to JPEG compression. The DCT itself is essentially lossless and fully reversible; however, the coefficients are rounded in a subsequent quantization stage. Quantizing all values obtained by the DCT is intended to achieve further, but lossy, data reduction. This is done by reducing the precision of the data through division and rounding, usually to whole numbers. A quantization table the size of the DCT matrix (8×8) provides an 8-bit encoded integer entry as a divisor for each DCT coefficient [70, 72]. This allows each DCT coefficient to be set and weighted individually (weighting matrix, Fig. 6). A smaller weight means a larger divisor [60]. The compression ratio can thus be adjusted at the expense of image quality.

> Compression methods from other media types can also be applied to video.

The compression methods presented here now also form the basis for various video formats, each of which uses one or more of these methods. Some selected representatives will be introduced in the next subsection.

2.4.2.1 Video Formats

Video formats refer to various recording methods for videos. The formats can be roughly divided into two categories: videotape formats (magnetic recording formats, MAZ formats) and file-based formats. File-based formats encode video information digitally, while videotape formats store information either analog or digital. Occasionally, television formats (HDTV, NTSC, PAL, SECAM, DVB, ATSC, ISDB) are also subsumed under video formats. Analog video storage is not covered in this book; instead, the following sections focus on the central digital formats. Here, too, it is important to be aware of the historical dimensions, as many of today's formats have emerged through continuous extensions of existing standards [68].

The MPEG (Moving Picture Experts Group) is a group of experts dedicated to the standardization of video compression and related areas such as audio data compression and container formats. In common usage, MPEG usually refers not to the expert group but to a specific MPEG standard. The MPEG meets three or four times a year for five-day meetings. About 350 experts from 200 companies and organizations from 20 different countries participate in these meetings, known as MPEG meetings. The official designation for MPEG is ISO/IEC JTC1/SC29/WG11 (International Organization for Standardization/International Electrotechnical Commission, Joint Technical Committee 1, Subcommittee 29, Working Group 11). Ideas from the International Telecommunication Union (ITU) have often been adopted, extended, and improved. An important basis for most modern (and now largely standardized) video formats is the **MPEG-1 standard** named after the group. This was developed in the 1980s (introduced in 1991) with the goal of compressing movies to the limited data rate (up to 1.5 Mbit/s) of an audio CD played at normal speed. The result, with correspondingly modest quality, is called Video CD. The data compression of a single image or a video sequence can be differentiated into four consecutive steps: image preparation, image processing, quantization, and entropy coding [69, 71, 72].

In **image preparation**, the structure of an image is defined to create a suitable digital representation of the image information. In MPEG, an image must be in YCrCb format, similar to the YUV format, consisting of three planes: the luminance component Y and the chrominance components Cr and Cb. The luminance component has twice the resolution of the others and, at the time of MPEG-1, could not exceed 768×576 pixels. Each pixel is encoded per plane with a depth of 8 bits. In addition to the image information, MPEG also stores the aspect ratio (1:1, 4:3, 16:9) and the frame rate (e.g., 25 Hz and 29.97 Hz). To enable effective motion compensation in later steps, which is usually only meaningful for partial areas, the image is again divided into "macroblocks." Such an MPEG macroblock consists of a 16×16 luminance component and two 8×8 color difference components, which are sequenced into six 8×8 blocks [71].

In **image processing**, the first compression methods are applied in preparation for quantization, such as DPCM and DCT [60]. The combination of DPCM and DCT is called hybrid DCT. Here, the DCT exploits spatial relationships and similarities between pixels within an image, while DPCM exploits similarities between neighboring images [70]. In MPEG, four different types of image coding must be distinguished, which differ in terms of image processing, quantization, and entropy coding [69].

- I-frames (Intra-Coded Pictures) are encoded without reference to information from other frames and serve as key frames for random access and fast forward. A DCT, as in JPEG, is applied to the 8×8 blocks defined in the macroblocks, and the resulting DC coefficients are encoded using intraframe DPCM. The differences are formed between consecutive blocks of a macroblock and then VLC encoded. The AC coefficients are encoded using RLC and VLC [69].

- P-frames (Predictive-Coded Pictures) are encoded and decoded based on information from preceding I- and P-frames, enabling a much higher compression rate compared to I-frames. However, to decode a P-frame, all preceding I- and P-frames must have been decoded [69]. The encoder determines a prediction frame from the previous one by decoding it through quantization by weighting with reciprocals and by the IDCT. The DPCM difference is formed from the prediction frame and the current frame. Prediction in MPEG is optimized by motion compensation [70]. For this, the encoder attempts to find a macroblock in the previous I- or P-frame that is as similar as possible to a block in the current frame (block matching). The minimum value of the sum of absolute differences of the luminance component yields the most similar macroblock. Only the motion vector (spatial difference of the macroblocks) and the difference values of all 8×8 blocks, each from the most similar and the macroblock to be encoded, are then encoded. The DCT is applied to these difference value blocks, and only those blocks that contain exclusively zero-valued coefficients are not further processed. After quantization, RLC and a Huffman-like VLC are applied. Due to the small differences in the motion vectors of neighboring macroblocks, these are encoded using DPCM and VLC [69].
- For encoding and decoding B-frames (Bidirectionally Predictive-Coded Pictures), information from previous and subsequent I- or P-frames is used, from whose macroblocks those of the current frame are generated. The two similar macroblocks can be interpolated by prediction, forming two motion vectors. A difference is then formed between the interpolated and the macroblock to be encoded. Entropy coding of the macroblocks is performed as in the P-frame. However, B-frames cannot be used as references for other frames [69]. Bidirectional prediction in MPEG allows a B-frame to be encoded with half the data of a P-frame [70].
- D-frames (DC-Coded Pictures) are used only for fast forward and are encoded as single frames. However, only the low-frequency components of a frame calculated by the DCT are encoded. They also consist of only one type of macroblock [71]. MPEG defines 8-bit quantization for luminance and chrominance components and optional standard quantization tables [70]. Quantization also differs between frame types, as the AC coefficients in I-frames are very small, while those in P- and B-frames are large [69].

The video data stream is organized in a six-layer hierarchy. The Sequence Layer is the top layer and manages the buffering of data. It includes the constant bit rate for a sequence as well as the minimum storage space required for decoding. The next layer, the Group of Pictures Layer (GOP Layer), contains cyclically recurring groups of several frames. The order of transmitted and displayed frames differs at this layer. An I-frame is always transmitted first so that the decoder can decode and store the reference frame. In the data stream, P-frames then follow as references for the subsequently transmitted B-frames [71]. No fixed number of P- and B-frames is defined for the GOP structure; however, the I-, P-, and B-frames repeat periodically according to the chosen intervals between them [70]. The third

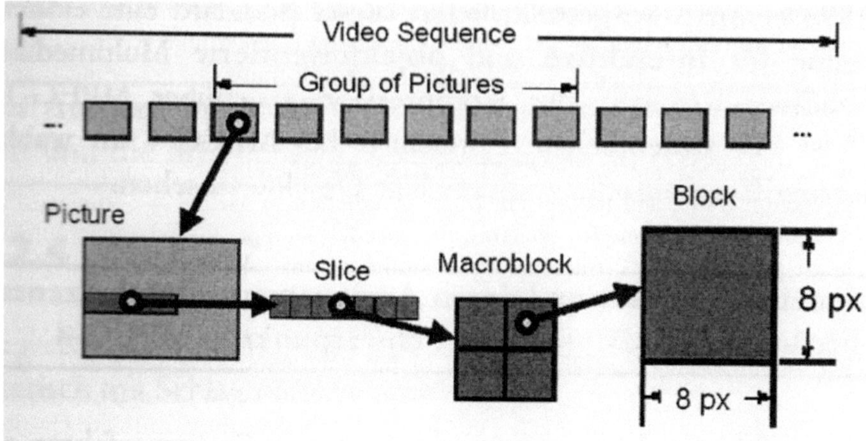

Fig. 2.39   The layers of MPEG-1

layer is called the Picture Layer. It contains only a single frame, whose temporal reference is defined by frame numbers. The next lower layer is the Slice Layer, which contains a slice composed of several macroblocks that may change in consecutive frames. Below this are the Macroblock Layer, Block Layer, macro- and 8×8 blocks [69]. The aggregation of the individual elements into a common data stream (multiplexing) is specified in the system definition. This also includes coordination of the data transfer of incoming and outgoing streams, buffer management, and clock synchronization. The data stream defined by ISO 11172 is divided into individual packs and provides the decoder with all necessary data, such as the maximum data rate of a pack. In an ISO-11172 data stream, header information is transmitted only at the beginning. MPEG also provides the timestamps required by such a data stream for synchronization [71] (see Fig. 2.39)

There are various MPEG formats with different purposes and areas of application.

MPEG-2 builds on the MPEG-1 standard as a backward-compatible extension and specifies high-quality video with a data rate of up to 100 Mbit/s. Extensions include, among others, a higher compression factor, higher image resolution, interlaced video formats, and HDTV support. MPEG-2 was structured according to application classes and defined as a set of extensible profiles. These profiles characterize the methods used for video coding. The Main Profile, for example, defines a data rate of about 2 to 80 Mbit/s and is intended for use in digital video transmission via cable and satellite, digital storage, and other communication applications. Like the High Profile, it can be used for HDTV applications. The Main Profile was extended by a hierarchical or scalable profile to enable applications such as terrestrial TV to maintain backward compatibility with other standards (MPEG-1, H.261) and to support applications requiring multi-layer coding. In addition to the essential extensions from MPEG-1 to MPEG-2, the scalability options for encoded moving images are noteworthy. When decoding the same data, several alternatives are available due to the encoding of different qualities. Spatial scalability allows image sequences of different resolutions to be decoded, with the luminance plane still encoded at full resolution and the chrominance planes subsampled by a factor of two. The number of full frames per second during playback is varied by scaling the frame rate. Amplitude scalability allows different bit depths or resolutions during quantization of the DCT coefficients, resulting in "layered coding" [69]. MPEG-2 supports the processing of interlaced fields, so encoding can be performed in either full-frame or field mode. When the image is static, similarities between neighboring lines of two fields can be exploited in full-frame mode. A full frame consists of 8×8 image blocks originating from both fields. In the case of significant motion in the block area, it is preferable to use field mode, as the fields differ greatly from each other [70, 71].

MPEG-4 is based, in terms of video coding, on the ITU standard H.263'9. The standard is organized into parts. It has been extended by Part 10, called MPEG-4/AVC (Advanced Video Coding), which was developed jointly with the ITU and is designated H.264'10 there. MPEG-4 offers improved audiovisual quality at lower data rates and improved coding efficiency; it also enables interactive handling of content through the introduction of an object-based coding model. To maintain backward compatibility, the basic coding described above is retained [70]. Innovations in MPEG-4 include "model-based image coding of human interaction with multimedia environments" [19] and speech coding at low data rates. The integration of various audio and video data types is also supported, including multiple camera perspectives (multiview video). MPEG-4 is retroactively extensible to adapt to advances in technology and methodology [69]. The syntactic description language MDSL11 is used to implement this concept [69]. To ensure stability in error-prone environments, MPEG-4 offers tolerance to residual errors, which in practice is realized as selective forward error correction, error containment, or error masking [45]. MPEG-4 provides not only playback but also interaction with content. It allows random access to video and audio scenes and editing of content. To enable the latter, audiovisual objects (AVO) were introduced. Users are given the ability to modify the content of a scene if it is composed of audiovisual objects. They can influence the shape, movement, texture, and position, access only parts of a sequence, or insert objects from one scene into another. A scene is thus a composition of these audiovisual data forms [69]. Further innovations in MPEG-4 include content-based manipulation and the ability to edit the bitstream without transcoding, which forms the basis for interactivity. Also introduced was hybrid coding and the combination of conventional or natural (e.g., live-action) and synthetic (e.g., text, graphics, sounds) audio and video data ("Synthetic Natural Hybrid Coding," SNHC) [73]. The decoder is given control over mixing these data. MPEG-4 extends the MPEG-1 coding scheme with an algorithm for shape coding and with shape-adaptive DCT ("Shape Adaptive DCT—SA-DCT"). The latter is an advanced version of the hybrid DCT used in MPEG-1 and 2, allowing transform coding of arbitrarily shaped image blocks. After transmitting the shape information for each object layer, the motion vectors of the macroblocks follow (macroblocks and blocks are split as in MPEG-1/2), as well as the DCT coefficients, which are associated with the motion and texture information (luminance and chrominance) of the respective object. In MPEG-4, different algorithms for temporal prediction can be applied to layers with different properties [73].

MPEG-7 occupies a special position among video formats, as it is—not strictly speaking—an actual video format, but rather a container format in which video content can be described with metadata [74–76].

```
 1  <Mpeg7>
 2      <Description xsi:type="ContentEntityType">
 3      <MultimediaContent xsi:type="ImageType">
 4      <Image id="IMG1">
 5      <SpatialDecomposition>
 6      <StillRegion id="SR1">
 7          <Semantic>
 8          <Label><Name>Roosevelt</Name></Label>
 9          </Semantic>
10      </StillRegion>
11
12      <StillRegion id="SR2">
13          <TextAnnotation>
14              <KeywordAnnotation>
15              <Keyword>Churchill</Keyword>
16              </KeywordAnnotation>
17          </TextAnnotation>
18      </StillRegion>
19
20      <StillRegion id="SR3">
21          <Semantic>
22              <Definition>
23              <StructuredAnnotation>
24              <Who><Name>Stalin</Name></Who>
25              </StructuredAnnotation>
26              </Definition>
27          </Semantic>
28      </StillRegion>
29          ...
30  </Mpeg7>
```

Listing 2.4 MPEG7 description of the image shown in Fig. 2.40

Listing 2.4 provides an example of how an image (see Fig. 2.40) can be annotated using MPEG-7. Such annotations (also called decompositions) are available in MPEG-7 for all multimedia types and were originally intended to help standardize metadata processing. With MPEG-7, all properties, objects, content, structures, technical or semantic features from multimedia or multimedia sequences can be described. A typical decomposition of this kind is shown in Fig. 2.41.

Fig. 2.40   Example of an MPEG-7 description [77]

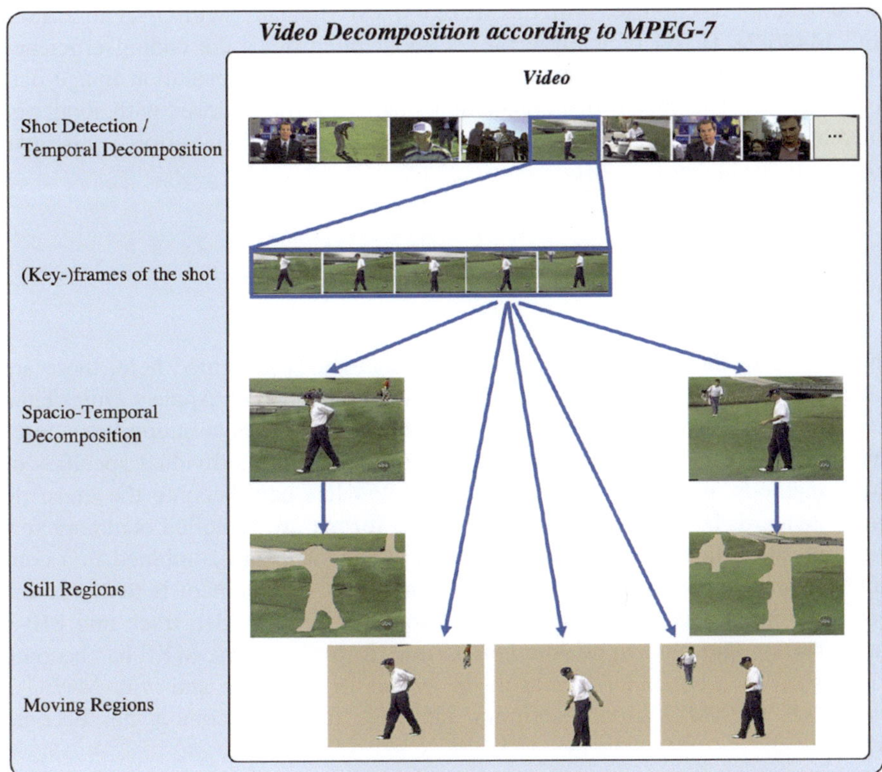

Fig. 2.41   Video decomposition with MPEG-7

Container formats contain additional metadata.

H.264/MPEG-4 AVC is now the de facto standard for video compression. It was initially developed by the ITU (Study Group 16, Video Coding Experts Group) under the name H.26 L. In 2001, the ITU group joined forces with MPEG-Visual and continued development jointly as the Joint Video Team (JVT). The goal of the project was to design a compression method that, compared to previous standards, would reduce the required data rate by at least half for the same quality, both for mobile applications and in TV and HD domains. Members of the working group include representatives from the Fraunhofer Institute for Telecommunications, Microsoft, and Cisco Systems. In 2003, the standard was adopted by both organizations with identical wording. The ITU designation is H.264. At ISO/IEC MPEG, the standard is known as MPEG-4/AVC (Advanced Video Coding) and is the tenth part of the MPEG-4 standard (MPEG-4/Part 10, ISO/IEC 14496-10). The standard should not be confused with the MPEG-4/ASP standard (MPEG-4/Part 2, ISO/IEC 14496-2). H.264 typically achieves about three times the coding efficiency of H.262 (MPEG-2) and, like H.262, is designed for high-resolution image data (e.g., HDTV). This means that comparable quality can be achieved with about one third of the MPEG-2 data volume. However, the computational effort is also two to three times higher [58, 78].

H.264 is currently the de facto standard.

In addition to the MPEG-based compression methods presented here, there are a number of other formats, such as Microsoft's AVI [79] or Apple's QuickTime format [80], each of which uses similar or sometimes identical internal structures and algorithms. However, this book does not go into the individual specifics of these methods, especially since MPEG-4/H.264 has now become the quasi de facto standard. Both AVI and the QuickTime format are so-called container formats, meaning that a number of video and audio tracks are combined in a container, which thus defines the type and structure of how content is to be stored. For example, an AVI container can contain an MPEG-4 video track and MP-3 audio tracks. The same applies to QuickTime. In recent years, MXF has become established as a container format in the professional sector, and with MPEG-7 and MXF, an XML-based description language for video content has become widespread.

2.5   **Extraction of Multimedia Features**

This subsection introduces a range of tools and APIs for extracting multimedia features. However, it should be noted right from the start that there is a vast array of different libraries and mechanisms for feature extraction from various multimedia formats, both in the open-source sector and in the commercial environment. The following selection is therefore intended merely to illustrate what feature extraction means in practice and what challenges developers have to face.

2.5.1   **Example Extraction of Text**

This subsection presents a concrete and practical Java example for extracting text from MS Office documents. Working with MS Office documents is a very widespread challenge. Simple document types, such as `.txt` or `.rtf`, can be processed directly and easily in the Java language environment. The same generally applies to `.xml` or `.json` files. However, when it comes to complex formats such as `.docx` or `.xlsx`, additional libraries are required. The most common representative for this is *Apache POI*[1] . With this library, all Microsoft Word document types can be created or read natively. Installation is done via download or *Maven*. The *Maven* coordinates are listed in Listing 2.5 [81].

```
1  <dependency>
2      <groupId>org.apache.poi</groupId>
3      <artifactId>poi-ooxml</artifactId>
4      <version>3.15</version>
5  </dependency>
```
Listing 2.5 Maven Coordinates for Apache POI

To provide an easy introduction to working with *Apache POI*, two examples are presented below: reading `.docx` documents and writing `.xlsx` documents.

Microsoft Word documents can embed not only a variety of formatting but also images or other multimedia content. This often makes extracting information quite complicated. Listing 2.6 shows an example that uses *Apache POI* to open a Word document, output its textual content, and save each embedded multimedia file as a separate file.

---

[1] https://poi.apache.org

```
 1 FileInputStream fis = new FileInputStream("Example.docx");
 2 XWPFDocument xdoc = new XWPFDocument(OPCPackage.open(fis));
 3 XWPFWordExtractor extractor = new XWPFWordExtractor(xdoc);
 4
 5 // Text part
 6 String content = extractor.getText();
 7
 8 StringTokenizer st = new StringTokenizer(content.toString(), "
      .");
 9 while (st.hasMoreTokens()) {
10     String sentence = st.nextToken();
11     System.out.println("Sentence:␣" + sentence);
12 }
13
14 // Multimedia part
15 ArrayList<PackagePart> parts = extractor.getPackage().getParts
      ();
16 for (PackagePart pp : parts) {
17     File fPart = new File("temp/" + pp.getPartName());
18     FileOutputStream fout = new FileOutputStream(fPart);
19     fout.write(pp.getInputStream().readAllBytes());
20     fout.close();
21 }
22 extractor.close();
```

Listing 2.6 Reading content out of Word documents

First, in line 1, a `FileInputStream` is created for the `.docx` file. Then, an `XWPFDocument` object is created, which represents the structure of the Word document and can be used to create an `XWPFWordExtractor` (line 3). The content of the document can then be read using the `getText()` method. In lines 9–12, this method is used to output the individual sentences of the document. Multimedia content is represented by `PackagePart` objects and can be read as shown in line 15. Each `PackagePart` here represents an embedded object, which can be processed using *streams*. In the example above, each object is saved as a file in a `temp` folder.

Of course, the *Apache POI* library has a large number of additional functions, but these would shift the focus of the book. Therefore, reference is made here to the vendor's website to explore further possibilities. In general, however, the examples in this subsection provide a very good demonstration of how to use *Apache POI* and serve as a solid entry point to this library.

Similar mechanisms for reading various file formats or the built-in functions for reading and writing files in Java can also be found in the book "Modern Software Development with Java and JEE" [82].

2.5.2 Example Extraction of Image Features

In the context of image feature extraction, the EXIF standard [83] plays a central role. It is considered one of the few industry-accepted and almost universally implemented standards. EXIF stands for "Exchangeable Image File Format" and stores important data for photos. These are embedded in the respective image formats (such as JPEG or TIFF) and can thus be used to enrich images with additional information, which is often used as key features. Many image editing programs offer editors for these EXIF data (see Fig. 2.42).

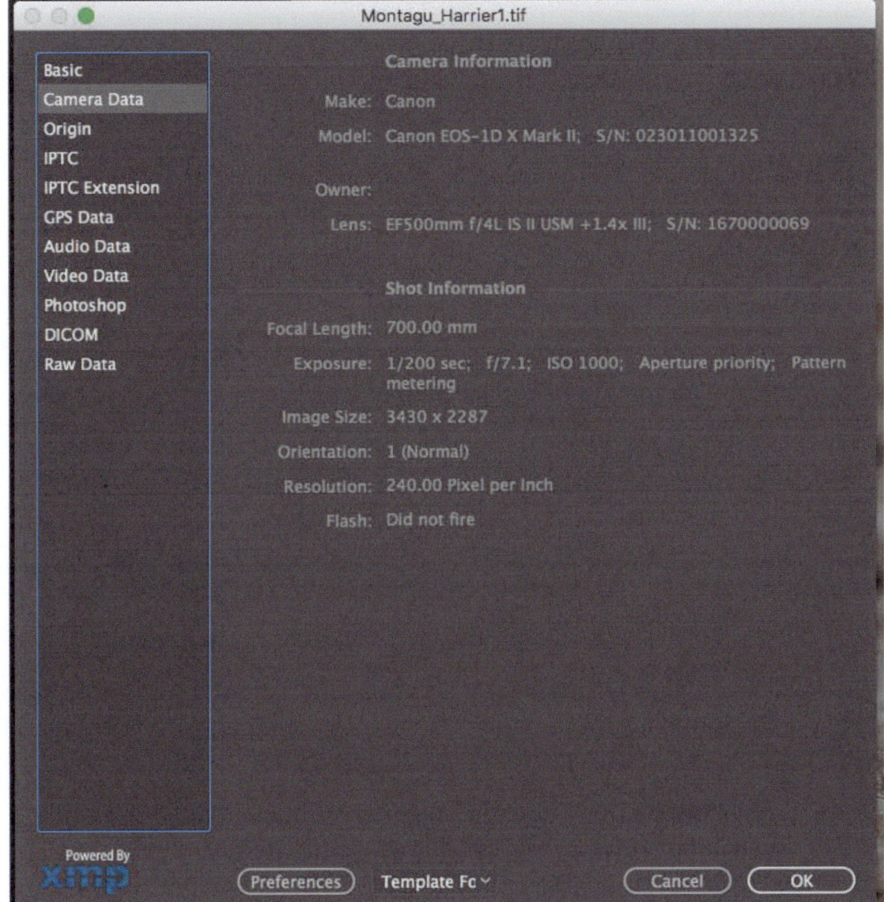

Fig. 2.42 EXIF viewer in Adobe Photoshop [5]

However, EXIF data can also be easily accessed programmatically. For this, the `commons.imaging` library from Apache can be used. The Maven coordinates can be found in Listing 2.7.

```
1  <dependency>
2      <groupId>org.apache.commons</groupId>
3      <artifactId>commons-imaging</artifactId>
4      <version>1.0</version>
5  </dependency>
```
Listing 2.7 Maven coordinates for Apache Commons Imaging

Using this library is quite straightforward and is shown in Listing 2.8. It is assumed here that the image is a JPEG. Exception handling is omitted to better illustrate the core of feature extraction.

```
1  File f = new File("some.jpeg");
2  ImageMetadata imd = Imaging.getMetadata(f);
3  JpegImageMetadata jpegMeta = (JpegImageMetadata)imd;
4  TiffImageMetadata exif = jpegMeta.getExif();
5  TiffImageMetadata.GPSInfo gpsInfo = exif.getGPS();
6
7  String exif_height = jpegMeta.findEXIFValue(TiffTagConstants.
      TIFF_TAG_IMAGE_LENGTH).getValueDescription();
8  String exif_width = jpegMeta.findEXIFValue(TiffTagConstants.
      TIFF_TAG_IMAGE_WIDTH).getValueDescription();
9  String exif_dpi = jpegMeta.findEXIFValue(TiffTagConstants.
      TIFF_TAG_IMAGE_WIDTH).getValueDescription();
10 String exif_aperture = jpegMeta.findEXIFValue(ExifTagConstants
      .EXIF_TAG_APERTURE_VALUE).getValueDescription();
11 String exif_brightness = jpegMeta.findEXIFValue(
      ExifTagConstants.EXIF_TAG_BRIGHTNESS_VALUE).
      getValueDescription();
12 String exif_shutterspeed = jpegMeta.findEXIFValue(
      ExifTagConstants.EXIF_TAG_SHUTTER_SPEED_VALUE).
      getValueDescription();
13 String exif_lens = jpegMeta.findEXIFValue(ExifTagConstants.
      EXIF_TAG_LENS_MODEL).getValueDescription();
14 String exif_camera = jpegMeta.findEXIFValue(ExifTagConstants.
      EXIF_TAG_BODY_SERIAL_NUMBER).getValueDescription();
15 String exif_focalLength = jpegMeta.findEXIFValue(
      ExifTagConstants.EXIF_TAG_FOCAL_LENGTH).
      getValueDescription();
16 String exif_iso = jpegMeta.findEXIFValue(ExifTagConstants.
      EXIF_TAG_ISO).getValueDescription();
17
18 String gpsDescription = gpsInfo.toString();
19 String exif_longitude = jpegMeta.findEXIFValue(GpsTagConstants
      .GPS_TAG_GPS_LONGITUDE).getValueDescription();
20 String exif_latitude = jpegMeta.findEXIFValue(GpsTagConstants.
      GPS_TAG_GPS_LATITUDE).getValueDescription();
```

Listing 2.8 Apache Commons Imaging Example

Here you can see that, based on a regular file (line 1), a number of objects can be instantiated, each containing different sets of metadata (lines 2–5). There are predefined field names as constants for the common EXIF data, which can be used to read the respective features (lines 7–20). This also shows how, for example, the camera used, geo-coordinates, or resolution information can be extracted. Each string variable thus holds the value for a specific feature and can be further processed accordingly. It should also be noted here that in this way we represent image features textually and can further process them using the mechanisms from Subsection 2.1. As already shown in Subsection 2.40, images can also be annotated and provided with metadata using MPEG-7. Since MPEG-7 is an XML format, such features can be extracted and further processed with any standard XML parser.

In addition to such standards, there are a whole range of APIs and services capable of extracting features from images. The entire discipline of pattern recognition [84] deals, for example, with recognizing patterns (and thus objects) in images. For further information, reference is made here to other works [84–86]. In the context of this book, the focus is not on developing algorithms that are particularly good at recognizing features, but rather on further processing these features within the framework of Multimedia Information Retrieval. Therefore, it is sufficient here to consider an example of such an API, which, behind the scenes, provides us with algorithms, machine learning methods [87], and a variety of other mechanisms according to best practices, ultimately delivering a set of features. As an example, we use Google Cloud Vision AI (coordinates see Listing 2.9)

```
1 <dependency>
2     <groupId>com.google.cloud</groupId>
3     <artifactId>google-cloud-aiplatform</artifactId>
4     <version>3.22.0</version>
5 </dependency>
```
Listing 2.9 Maven coordinates for Google Cloud AI

Google Vertex AI can be used for a variety of feature extraction tasks. Similar and comparable APIs are also available from Microsoft [88] and Amazon [89]. Their usage is also similar. It is also comparable that these three APIs are commercial services that must be licensed and paid for accordingly. For smaller projects, the respective accesses are free of charge. In the open-source environment, projects such as Yolo [90], Image AI [91], Clarifai [92], or Tensor Flow [93] could be mentioned. Listing 2.10 shows the use of Google Vertex AI for extracting recognized objects, Listing 2.11 for extracting dominant (i.e., frequently used) colors.

```
1 Image img = Image.newBuilder().setContent(imgBytes).build();
2 AnnotateImageRequest request =
3     AnnotateImageRequest.newBuilder()
4     .addFeatures(Feature.newBuilder().setType(Type.
         OBJECT_LOCALIZATION))
5     .setImage(img)
6     .build();
7 requests.add(request);
8
9 ImageAnnotatorClient client = ImageAnnotatorClient.create());
10 BatchAnnotateImagesResponse response = client.
      batchAnnotateImages(requests);
11 List<AnnotateImageResponse> responses = response.
      getResponsesList();
12
13 for (AnnotateImageResponse res : responses) {
14     for (LocalizedObjectAnnotation entity : res.
          getLocalizedObjectAnnotationsList()) {
15         System.out.format("Object_name:_
              %s%n", entity.getName());
16         System.out.format("Confidence:_
              %s%n", entity.getScore());
17     }
18 }
```
Listing 2.10 Usage of Google Vertex AI

```
1 List<AnnotateImageRequest> requests = new ArrayList<>();
2 ByteString imgBytes = ByteString.readFrom(new FileInputStream(
      filePath));
3
4 Image img = Image.newBuilder().setContent(imgBytes).build();
5 Feature feat = Feature.newBuilder().setType(Feature.Type.
      IMAGE_PROPERTIES).build();
6 AnnotateImageRequest request =
7     AnnotateImageRequest.newBuilder().addFeatures(feat).
          setImage(img).build();
8 requests.add(request);
9
10 ImageAnnotatorClient client = ImageAnnotatorClient.create());
11 BatchAnnotateImagesResponse response = client.
      batchAnnotateImages(requests);
12 List<AnnotateImageResponse> responses = response.
      getResponsesList();
13
14 for (AnnotateImageResponse res : responses) {
15     DominantColorsAnnotation colors = res.
          getImagePropertiesAnnotation().getDominantColors();
16     for (ColorInfo color : colors.getColorsList()) {
17         System.out.format(
18         "fraction:_%f%nr:_%f, _g: _%f, _b: _%f%n",
19         color.getPixelFraction(),
20         color.getColor().getRed(),
21         color.getColor().getGreen(),
22         color.getColor().getBlue());
23     }
24 }
```
Listing 2.11 Detection of dominant colours

In both listings, it is clear that the creation and access to the Google API are consistent. An `ImageAnnotatorClient` is instantiated in each case, which can receive various parameters specifying, for example, the type of feature extraction to be performed (Listing 2.10 line 4, Listing 2.11 line 5). It is also clear that the results depend on the expected feature. In Listing 2.10, a `LocalizedObjectAnnotation` is used for this purpose, in Listing 2.11 a `DominantColorsAnnotation`. In both cases, the identified features are represented textually.

2.5.3 Example Extraction of Audio Features

Of course, Google, Microsoft, and Amazon also offer solutions for extracting audio features. Therefore, an example of audio transcription based on Google Vision is shown first (see Listing 2.12).

```
1  LoadBalancerRegistry.getDefaultRegistry().register(new
       PickFirstLoadBalancerProvider());
2  RecognitionConfig recConfig = RecognitionConfig.newBuilder()
3      .setEncoding(AudioEncoding.ENCODING_UNSPECIFIED).
           setLanguageCode("en-US")
4      .setSampleRateHertz(16000).setModel("video").build();
5
6  RecognitionAudio recognitionAudio = RecognitionAudio.
       newBuilder().setContent(ByteString.copyFrom(content))
7  .build();
8
9  RecognizeResponse recognizeResponse = speech.recognize(
       recConfig, recognitionAudio);
10
11 List<SpeechRecognitionResult> list = recognizeResponse.
       getResultsList();
12 for (SpeechRecognitionResult result : list) {
13     SpeechRecognitionAlternative alternative = result.
           getAlternativesList().get(0);
14     System.out.printf("Transcript␣:␣
           %s\n", alternative.getTranscript());
15 }
```

Listing 2.12 Usage of Speech Recognition

This short example clearly demonstrates how easy it is to extract features using the appropriate APIs. After the API is instantiated and configured (lines 1–7), the actual recognition is called in line 8. The results then come as a list with individual segments/sentences and can be processed further in the same way as a text file (lines 12–15).

In addition to audio transcription, music recognition also plays an important role in practice. A very interesting and well-known example is Shazam by Apple **shazam.** Shazam can be used via a standardized REST interface and provides access to an extensive media library, which can be accessed using the query-by-example paradigm [1]. That is, a sample file is uploaded (e.g., a snippet from a music clip) and Shazam determines the corresponding title, artist, album, etc. The result of the web service call is a JSON file, an excerpt of which is shown in Listing 2.13.

```json
1  {
2  ...
3  "timezone":"Europe/Paris",
4  "track":{
5      "albumadamid":"380907762",
6      "genres":{ "primary":"Pop"},
7      "hub":{
8          "actions":[{
9              "name":"apple",
10             "type":"uri",
11             "uri":"https://audio-ssl.itunes.apple.com/itunes-
                   assets/AudioPreview112...18680689350248047.
                   plus.aac.ep.m4a"
12         }
13         ],
14         "displayname":"APPLE_MUSIC",
15         ...
16         "providers":[{
17             "actions":[{
18                 "name":"hub:spotify:searchdeeplink",
19                 "type":"uri",
20                 "uri":"spotify:search:Take%20On%20Me%20a-ha"
21             }],
22             "caption":"Open_in_Spotify",
23             "images":{
24                 "default":"https://images.shazam.com/static
                       /...spotify_{scalefactor}.png",
25                 "overflow":"https://images.shazam.com/static
                       /...spotify-overflow_{scalefactor}.png"
26             },
27             "type":"SPOTIFY"
28         },
29         ...
30     },
31     "images":{
32         "background":"https://is2-ssl.mzstatic.com/image/thumb
                   /...ghez.jpeg/800x800cc.jpg",
33         "coverart":"https://is3-ssl.mzstatic.com/image/thumb
                   /...cmc.jpg/400x400cc.jpg",
34         "coverarthq":"https://is3-ssl.mzstatic.com/image...
                   zfcmc.jpg/400x400cc.jpg",
35         "joecolor":"b:e3e2dep:0d0e0fs:222223t:383838q:484848"
36     },
37     "isrc":"USWB11001072",
38     "key":"60695704",
39     "layout":"5",
40     "sections":[{
41         "metadata":[{
42             "text":"Hunting_High_and_Low_(Deluxe Edition)",
43             "title":"Album"
44         },{
```

```
45              "text":"Rhino/Warner Records",
46              "title":"Label"
47          },{
48              "text":"1985",
49              "title":"Released"
50          }],
51          "tabname":"Song",
52          "type":"SONG"
53      },{
54          "beacondata":{
55              "commontrackid":"11738347",
56              "lyricsid":"29304649",
57              "providername":"musixmatch"
58          },
59          "footer":"Writer(s): Magne Furuholmen, Morten Harket,
                Paul Waaktaar-savoy\nLyrics powered by
                www.musixmatch.com",
60          "tabname":"Lyrics",
61          "text":[
62          "We're talking away",
63          "I don't know what I'm to say",
64          "I'll say it anyway",
65          "Today is another day to find you",
66          ...
67          ],
68          "type":"LYRICS"
69      }],
70      ...
71      "subtitle":"a-ha",
72      "title":"Take On Me",
73      "type":"MUSIC",
74      "url":"https://www.shazam.com/track/60695704/take-on-me",
75      "urlparams":{
76          "{trackartist}":"a-ha",
77          "{tracktitle}":"Take+On+Me"
78      }
79  }
80  }
```

Listing 2.13 Result of a Shazam-API-Call

This listing gives an initial impression of how extensive the extracted features can be in practice. For example, you will find the title of the uploaded music excerpt (line 72), the artist (line 76), links to photos of the album cover (line 33), the ISRC number (line 37), or even the complete song lyrics (line 62ff). It is now up to the subsequent process to determine from this multitude of extracted features those that are relevant for further processing.

2.5.4 Example Extraction of Video Features

Video basically consists of individual images and associated audio tracks. Therefore, all methods for feature extraction from images and audio signals can also be applied to video, provided the video can be split into individual frames or the audio track can be extracted. The greatest difficulty in practice is the many different video and audio formats, since in theory, feature extraction must be adapted for each of these formats. Fortunately, there are various tools that can help here. One of the most important is `FFmpeg` [94], which is almost considered the "Swiss Army knife" for video editing and conversion in the industry. Listing 2.14 shows some examples.

```
1 ffmpeg -i video.mp4
2 ffmpeg -i video.mp4 video.avi
3 ffmpeg -i video.mp4 -vn audio.mp3
4 ffmpeg -i video.mp4 -filter:v scale=1280:720 -c:a copy output.
      mp4
5 ffmpeg -i video.mp4 -vf fps=1 out%d.png
```
Listing 2.14 FFmpeg examples

In line 1, metadata for the video `video.mp4` is simply output as text. This includes all accessible metadata, ranging from technical metadata such as codec, resolution, etc., to textually annotated metadata such as copyright information. Line 2 converts the video to AVI format. Line 3 shows how the audio track of a video can be extracted and saved in mp3 format. Line 4 converts and scales a video to the format 1280×720, and line 5 shows how to extract one image per second from the video, named `out1.png, out2.png, ....` On this basis, videos can be brought into the correct format for any subsequent processes.

Nevertheless, it should also be mentioned here that the APIs already mentioned are, of course, also capable of processing videos directly and extracting corresponding features.

> There are well-usable libraries for metadata extraction.

2.6 Machine Learning and Artificial Intelligence

So-called artificial intelligence has now found its way into almost all areas of feature extraction. In the context of Multimedia Information Retrieval, there are naturally also areas of application and points of connection. However, it is first necessary to provide a clear definition of "artificial intelligence" that is also applicable in the field of MMIR. It turns out that the literature has not yet managed to agree on a description. The following examples illustrate this:

- Bitcom e. V. and German Research Center for Artificial Intelligence **bitcom**: "Artificial intelligence is the property of an IT system to exhibit human-like, intelligent behaviors."
- European Parliament [95]: "a machine-based system designed to operate with varying degrees of autonomy and, after its introduction, demonstrates adaptability, and that, for explicit or implicit objectives, derives from the inputs it receives how it can generate outputs such as predictions, content, recommendations, or decisions that can influence physical or virtual environments."
- Spektrum der Wissenschaft, Lexikon der Neurowissenschaften [96]: "Artificial intelligence [...] is a subfield of computer science that deals with the study of mechanisms of intelligent human behavior [...]".
- Microsoft Corp. [51]: "By artificial intelligence (AI), we mean technologies that complement and enhance human abilities in seeing, hearing, analyzing, deciding, and acting."

Even though the world of definitions is obviously still quite inconsistent, a basic classification of AI systems has nevertheless become established (see Fig. 2.43), which can also be used to distinguish and categorize Multimedia Information Retrieval systems. Algorithmic decision systems are generally already referred to as "weak AI". This is because simple tools, such as a spell checker, are already considered "intelligent" systems. Similarly, all MMIR systems also fall into this category, as they are able to manage and present knowledge and information to the user in an "intelligent" way on demand. This area mainly includes static, non-learning systems, such as knowledge-based systems (including MMIR systems), rule-based systems, and expert systems. In the area of dynamic and learning systems, the discipline of machine learning (ML) is particularly noteworthy. ML systems are used to train neural networks based on existing information, which can

Classification of KL systems

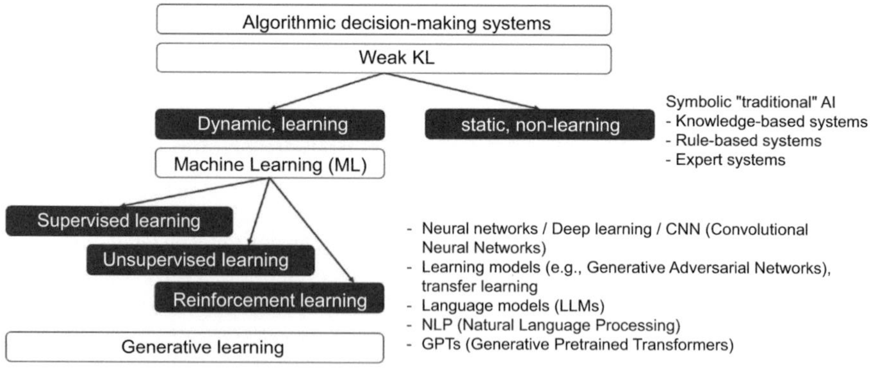

Fig. 2.43 A possible classification of AI systems

then be applied automatically to link information and solve various tasks. Typical areas of application in MMIR include object recognition, feature extraction, quality control in digital signal processing, and pattern recognition in large data sets (big data). Such methods can be divided into three subcategories [29, 87]:

- **Supervised learning:** Here, the learning process is continuously monitored by a user. That is, if, for example, daisies are to be recognized in an image, a user would manually check each image given to the ML method for learning and provide immediate feedback.
- **Unsupervised learning:** In this approach, there is a set of manually prepared training data with positive and negative cases (essentially daisies and not daisies). These are then automatically passed to the learning method with the expectation that similar data can also be processed correctly in the future. The quality of such methods naturally increases with the quality and scope of the training data.
- **Reinforcement learning:** Here, two AIs are either made to cooperate or compete against each other. For example, one AI could be taught to generate photos of daisies and the other to recognize daisies in photos. The idea behind reinforcement learning is that the combination of the two AIs leads to both essentially training each other, thereby improving the quality of each one.

Applications from the field of these methods have given rise to a new discipline, so-called generative AI, which is capable of actively producing information. Examples include Chat-GPT [4] for natural language information processing or Midjourney [97] or Dall-E [98] for generating images as examples for queries (so-called prompts). Generative AI creates new information from the (extremely extensive) training data.

The example in Fig. 2.44 also highlights an important difference between generative AI systems and MMIR systems: While MMIR systems focus on finding suitable matches to a query such as "a contemporary image of a professor talking about digital art in a podcast" from the information base using the provided features—in other words, the result is information (or images) that actually exist and were at some point loaded into the information base and processed by the MMIR system—generative AI would not even "bother" to search for images, but would instead simply draw a suitable image. The resulting questions regarding rights, accuracy, or ethical frameworks are currently being hotly debated in the media. However, this aspect will not be discussed further in this book.

> Machine learning is frequently used for feature extraction.

Whereas machine learning methods are extremely frequently used for feature extraction in MMIR applications, generative AI methods contradict the fundamental nature of MMIR applications. As part of an outlook and a more detailed

Fig. 2.44 Generative AI using the example of Dall-E, the corresponding prompt is shown on the right [99]

distinction, the topic of generative AI will be revisited and examined in more detail in Subsection 6.2 at the end of the book.

2.7   Representation

So far, we have mainly dealt with the textual representation of individual features and have seen that this exists and can be used for all multimedia types. However, it is often the case that a multimedia object contains not just a single feature to be extracted, but usually a whole series of features at different levels. As already shown in Example 2.3, for instance, the extraction of audio features can yield hundreds of different features, each with its own textual representation. These features and their representations are often organized in so-called feature graphs [100].

Typically, such graphs are constructed in MMIR applications and enriched with features through the processing of multimedia objects. A graph G can be represented by $G = (N, E)$, where N is a set of nodes and E is a set of edges (sometimes also called relationships). If nodes and/or edges consist of multimedia features, G can also be called a feature graph or multimedia graph. Such feature graphs can be structured in various ways; it is important to distinguish what semantics—that is, meaning—the nodes and edges are assigned, and thus, what structure results. Typically, however, feature graphs are modeled as directed graphs and managed and organized in this form within the MMIR system. Whenever new features are detected, algorithms or techniques are needed to insert these new features into existing feature graphs and to ensure that the technical and semantic structures of feature graphs remain valid. Existing XML formats, such as MPEG-7, inherently possess graph-based structures, since any XML document can also be represented as a directed graph [101, 102].

Graphs represent a simple and extremely flexible data structure whose structure and size can vary depending on the application. Regardless of the information they represent, a number of algorithms are, of course, required for working with graphs. Typical examples include

- finding the shortest connections between two nodes,
- finding similar structures,
- moving, inserting, and deleting nodes,
- calculating edge weights
- and much more.

Depending on the size of the graphs, such operations can require very long computation times and thus also long runtimes. Storing, managing, and algorithmically processing graphs therefore presents a major challenge, which is addressed, for example, by graph databases. Such databases provide a generic storage structure that enables specialized query languages and optimized processing. The most common representative is Neo4J [103]. This graph database implements a number of algorithms that can be applied using a special query language, Graph-QL [104]. Listing 2.15, for example, shows how similarity search can be applied to a graph.

```
1  MATCH (r1:Node {name: 'N_Root_Image_Search'})-[*..4]->(child1)
2      WITH r1, collect(id(child1)) AS r1Child
3      MATCH (i:Image)-[:img]->
4          (r2:Node {name: 'N_Root_Image'})-[*..4]->(child2)
5          WHERE r1 <> r2
6          WITH r1, r1Child, r2, collect(id(child2)) AS r2Child
7
8  RETURN r1.image AS from, r2.image AS to,
9      gds.alpha.similarity.jaccard(r1Child, r2Child)
10     AS similarity
```
Listing 2.15 Ähnlichkeitssuche in Neo4J

In general, graphs are simple and well-studied mathematical objects that can be represented in various ways. Since a number of operations on feature graphs are necessary, the following introduces the mathematical foundations. Essentially, graphs are transformed by so-called encoders into other mathematical spaces in which certain computations can be performed more easily or quickly. The reverse conversion (if possible) is performed by a decoder [105]. Here, functions f_{Enc} for encoding and f_{Dec} for decoding a d-dimensional space ($d < |N|$) can generally be described as follows:

$$f_{\text{Enc}} : N \to \mathbb{R}^d$$
$$f_{\text{Dec}} : \mathbb{R}^d \times \mathbb{R}^d \to \mathbb{R}^+$$

The encoder function f_{Enc} maps individual nodes of the graph to elements $z_i \in \mathbb{R}^d$, where z_i represents the so-called embedding for the node $n_i \in N$. The decoder function f_{Dec} is typically specified as a pairwise decoder, which maps a pair of embeddings (z_i, Z_j) to a proximity measure P_G, referring to the original node pair (n_i, n_j) of the original graph. The proximity measure P_G thus also indicates the quality of the reconstruction of the decoding function f_{Dec}:

$$f_{\text{Dec}}((f_{\text{Enc}}(n_i), f_{\text{Enc}}(n_j)) = f_{\text{Dec}}(z_i, z_j) \approx P_G(n_i, n_j)$$

This general definition does not require that f_{Enc} and f_{Dec} be fully convertible into each other or losslessly converted. There may (and will) be a number of algorithms that use f_{Enc} to transform graphs into very restricted spaces and possibly even compress them. To describe precisely this deviation between the decoded proximity value and the actual $P_G(n_i, n_j)$, a loss function f_{Loss} is typically introduced:

$$P_G(n_i, n_j) = f_{\text{Loss}} \cdot f_{\text{Dec}}(z_i, z_j)$$

Based on $f_{\text{Enc}}, f_{\text{Dec}}, P_G$ and f_{Loss}, there are now a number of typical graph encodings [100, 105, 106]:

- Approaches for direct encoding: $f_{\text{Enc}}(n_i) = Zn_i$, where Z is a matrix containing all embeddings of all nodes.
- Factorization: These approaches are similar to classical matrix factorization and reduce the number of dimensions. A good example is the Laplacian eigenbasis, in which

$$f_{\text{Dec}}(n_i, n_j) = \|n_i - n_j\|_2^2$$

In this case, the quality Q can be calculated as follows

$$L = \sum_{(n_i, n_j) \in N} f_{\text{Dec}}(z_i, z_j) \cdot f_{\text{Loss}}(n_i, n_j)$$

- So-called random walk approaches: Based on direct encoding, the embeddings are optimized so that nodes receive similar embeddings if they are connected by short, random paths in the graph. This stochastic method achieves a number of excellent results in various fields [107].
- Deep Walk and node2vec: These are methods that apply machine learning to optimize the encoding of nodes using learning algorithms based on short, random walks.

> Feature graphs can represent multiple features in context.

From a mathematical perspective, a graph G with n nodes can also be represented by its adjacency matrix AM [108]. AM contains one row and one column per node, thus creating a matrix with n rows and n columns. If there is an edge between two nodes n_1, n_2, then the matrix position (n_1, n_2) is assigned the value 1, indicating that an edge exists between the two nodes. Fields with the value 0 indicate that there is no edge between the respective nodes. Figure 2.45 shows an example graph G_{ex}, which consists of five nodes and six edges.

The corresponding adjacency matrix of this graph $AM(G_{\text{ex}})$ is as follows:

Fig. 2.45 A simple graph G_{ex}

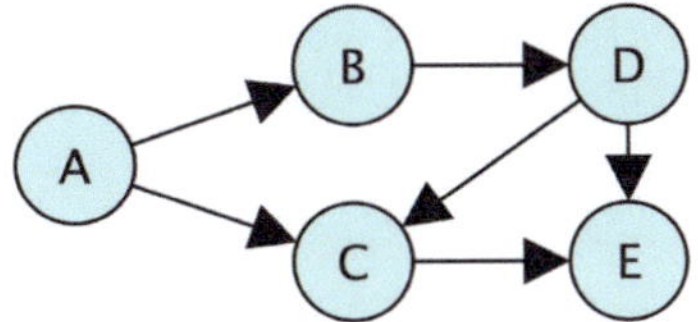

$$AM(G_{ex}) = \begin{pmatrix} 0 & 0 & 0 & 0 & 0 \\ 1 & 0 & 0 & 0 & 0 \\ 1 & 0 & 0 & 1 & 0 \\ 0 & 1 & 0 & 0 & 0 \\ 0 & 0 & 1 & 1 & 0 \end{pmatrix}$$

In $AM(G_{\text{ex}})$, the rows and columns $\{A, B, C, D, E\}$ are implicitly assumed. For directed graphs, AM is asymmetric; for undirected or fully bidirectionally directed graphs, AM is symmetric. Often, instead of the value 1 for representing an edge, other values are entered in AM, for example, for the length of an edge or the so-called weight. For weighted graphs, values other than 0 and 1 are therefore possible in AM. Another difference lies in the configuration of the diagonal. Depending on f_{Enc}, the diagonal can be filled with 0, 1, or any other value. For example, if nodes and edges have different types, the values of AM can also represent a corresponding type. Figure 2.46 shows a simple feature graph with different types.

The corresponding adjacency matrix $AM(G_{\text{ex2}})$ for this simple graph is as follows, assuming the following type coding:

- n: node, blue color, type $= 11$
- s: synonym, green color, type $= 12$
- cn: child node, blue color, type $= 21$
- sr: spatial property, type $= 22$
- s: synonym edge, type $= 23$

What is important here (and especially for further mathematical operations) is the order in which the rows and columns are arranged. A clearly defined order enables fast comparisons, sorting, or searches in the vocabulary of a feature graph via the

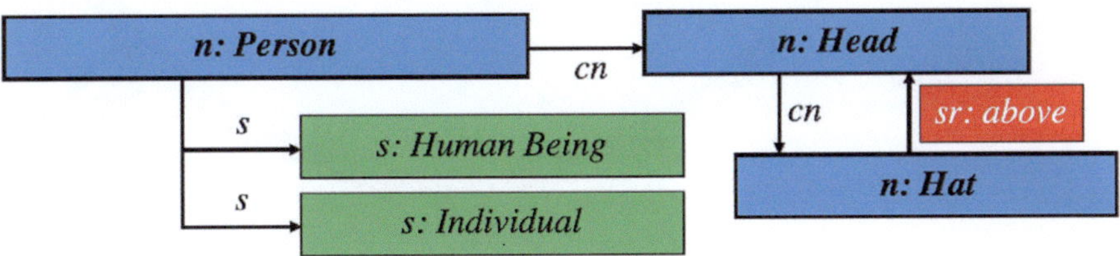

Fig. 2.46 A simple feature graph

corresponding adjacency matrix. For this example, we assume the following order: [Person, Head, Hat, Human Being, Individual].

$$AM(G_{\text{ex2}}) = \begin{pmatrix} 11 & 21 & 0 & 23 & 23 \\ 0 & 11 & 21 & 0 & 0 \\ 0 & 22 & 11 & 0 & 0 \\ 0 & 0 & 0 & 12 & 0 \\ 0 & 0 & 0 & 0 & 12 \end{pmatrix}$$

This example also shows that the coding for the spatial property would need to be extended in practice. Here, further sub-codings could be introduced, for example, for the terms "beside", "below", "in front of", "behind", etc., each of which would need its own value for the matrix coding. All of this is part of the function f_{Enc}. In addition to these considerations, there are a number of further optimizations in practice for the mapping rules of graphs into other mathematical spaces. However, for the purposes of this book, the current understanding is already sufficient to enable the concrete application of feature graphs as index structures.

Due to the ever-increasing level of detail and ever-higher resolution of multimedia objects, feature graphs in practice will quickly become very large (see Fig. 2.47). This negatively affects processing time and necessitates the use of optimized methods and algorithms, such as the use of matrix-based computations.

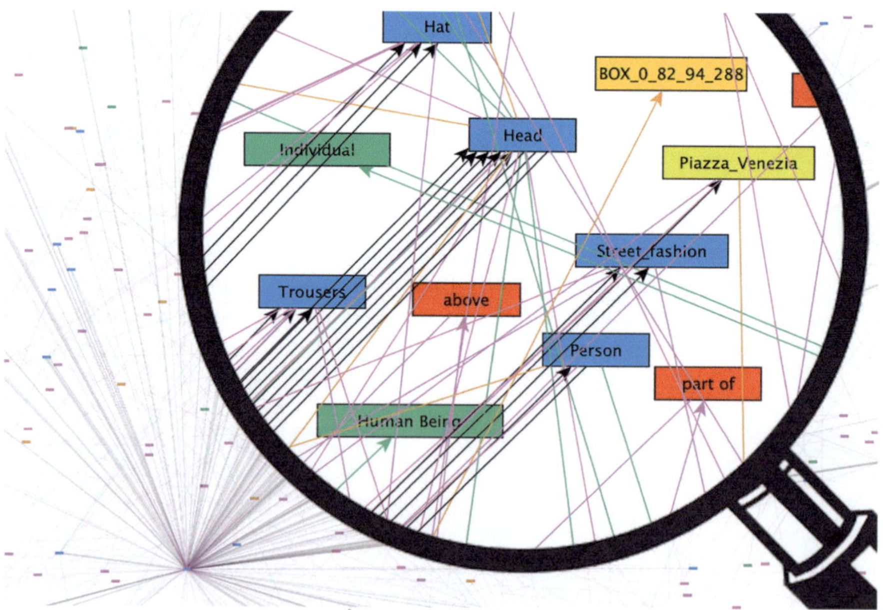

Fig. 2.47 A fully developed feature graph

So far, the structure of feature graphs and the representation of features have been explained. However, textual identifiers have always been used for the features or their values. Textual identifiers are, of course, subject to the problem that users may have made spelling mistakes when annotating features or creating texts, that different languages may need to be reconciled, that synonyms must be resolved, and that the meaning of terms may need to be specified more precisely. Such problems can usually only be solved by incorporating semantics. What this means will be explained in the next subsection.

2.8 Semantics

By introducing semantics, the previously rather purely alphabetical representation of features can be extended by their meaning. A simple, often-cited, but still illustrative example is the term "Jaguar." This term can be used to describe an animal, a car brand, or a company (and not many other things, such as an avatar in a virtual world), and based solely on the sequence of letters, it is not possible to determine in which context the term is being used. Even if it is clear, for example, that it refers to the animal, this can in turn be described in different languages using different character sequences. While the Finnish term "jaguaari" is still somewhat similar to the German and could possibly be related through a stem analysis (see Subsection 2.7), the characters of languages such as Bulgarian, Georgian, Russian, or even Chinese have nothing to do with the textual representation we are accustomed to. From a systems perspective, this means that the textual representation of features is actually only of very limited use as a universal representation.

Semantics attempts at this point to abstract away from any textual representation and instead focus on meaning. The meaning is defined in the Duden [109] as: "the sense inherent in actions, circumstances, things, phenomena" or "the conceptual content of a sign; the relationship between the word form and the conceptual content." Semantics [109] in turn is defined as: "a subfield of linguistics that deals with the meanings of linguistic signs and sequences of signs." From these definitions, it is clear that semantics makes use of the concept of meaning and focuses on working with language and text. Humans grasp the meaning of a document's content through their learned understanding of the world. Machines, however, cannot achieve this understanding of the meaning of a document's content without further intervention. For example, as humans, we can intuitively and clearly assign the sentence "The jaguar is now rarely found in the dense jungles of South America." to a biological topic, based on the context of the sentence. We can also clearly distinguish it from the car brand "Jaguar." A computer, however, requires additional information or an explicit, machine-understandable classification of the situation to make this distinction. The basis for this is provided by so-called **classifications**, which—generally formulated—are a method for determining membership. While classifications are still quite general and unstructured, **taxonomies** already provide a clear and systematic framework by defining an order of classes within a domain of knowledge. The term originally comes from the animal and

Fig. 2.48 Taxonomic classification of the jaguar [112]

plant world and was used for the clear identification of individual species and the depiction of their relationships to one another. Typically, taxonomies use the "is-a" relationship and thus form hierarchical structures. Figure 2.48 shows the taxonomic classification of the jaguar [110, 111].

Taxonomies map the "is-a" relation.

Of course, the "is-a" relationship and a taxonomy grid predefined for a given subject area are often inflexible. Therefore, an extension in the form of a **thesaurus** is often used. This is a controlled vocabulary, i.e., predefined terms specify which concepts are available for describing relationships. Terminological control in this context means that ambiguities and vagueness of natural language are to be reduced. This is achieved, on the one hand, by grouping synonyms (i.e., different terms with the same meaning) and alternative spellings into equivalence classes. Ambiguities arising from homographs (terms with the same spelling but different meaning and pronunciation—e.g., tenor: attitude vs. vocal range) and polysemes (terms with different meanings—bank: financial institution vs. bench) are resolved by splitting them into multiple equivalence classes. It is also usually the case that one element from each equivalence class is selected as the preferred element and then used to name the class—this preferred element is called the descriptor. In addition to the hierarchical relationships between broader and narrower terms, descriptors in a thesaurus can also refer to other related descriptors (association relation) [110].

> Thesauri are controlled vocabularies.

Thesauri also exist for various subject areas. Figure 2.49 shows, for example, the Agrovoc Multilingual Thesaurus [113], which addresses the above-mentioned problem of multilingualism and includes translations in all major languages in the controlled vocabulary. The aspect of taxonomy is also taken into account, as the individual terms can also be arranged hierarchically. All these methods are used for the textual and semantic representation of terms (and, in the context of MMIR, also features). In recent years, however, the emergence of the Semantic Web [114] has introduced another comprehensive concept, which will be presented in the next Subsection.

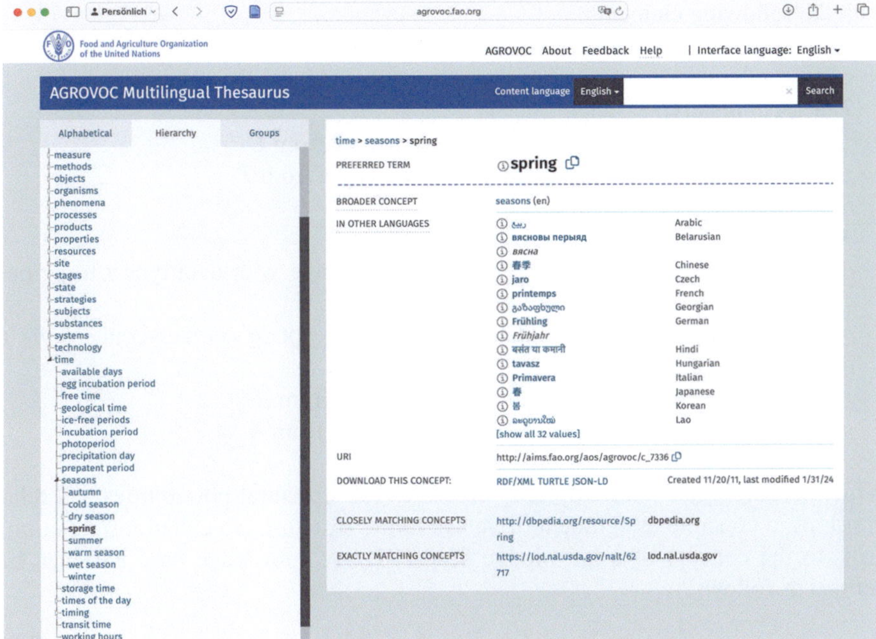

Fig. 2.49 Example of a thesaurus [113]

2.8.1 Ontologies and the Semantic Web

The term ontology (from the Greek "on," "being," and "logos," "doctrine") originally comes from philosophy [115]. Ontology is one of the oldest philosophical disciplines and deals with being itself and its organization. In information science, ontologies are, for example, defined as follows [116]: "An ontology is an explicit specification of a conceptualization. The term is borrowed from philosophy, where an Ontology is a systematic account of Existence." At its core, it is about creating a conceptualization, which according to Perez [117] can have a number of different components:

- **Concept:** the representation of an object of consideration to be described (this can be anything in the broadest sense, e.g., people, documents, processes, or similar),
- **Relation:** the specification of the connection between concepts,
- **Function:** relations with constraints (e.g., the relation motherOf, where, for example, each child can have only one mother),
- **Instance:** correspond to the elements of an ontology (value or instantiation of a concept or a relation).

Formally, ontologies can be described according to the definition of Ehring [118]: an ontology O_T is a structure

$$O_T := \{C, T, \leq_C, \leq_T, R, A, \rho_A, \rho_R, \leq_R, \leq_A\} \tag{2.8}$$

with the following elements:

- C: a set of concepts
- T: a set of data types
- $\leq_C$: the hierarchical arrangement of the concepts from C
- $\leq_T$: the hierarchical arrangement of the data types from T
- R: a set of relations
- A: a set of attributes
- ρ_A: an assignment function that assigns an attribute with data type t to a concept: $A \rightarrow C \times T$
- ρ_R: an assignment function that connects a concept to another concept via a relation $R \rightarrow C \times C$
- $\leq_R$: the hierarchical arrangement of the relations from R
- $\leq_A$: the hierarchical arrangement of the attributes from A

While an ontology O_T initially only describes the structural possibilities and relationships, a concrete assignment of values to the structure (so-called instantiation) leads to the concept of a knowledge base (KB). A knowledge base KB_T can be defined as follows:

$$KB_T := \{C_{KB}, T_{KB}, R_{KB}, A_{KB}, I, V, i_C, i_T, i_R, i_A\} \tag{2.9}$$

The first four parameters are assigned as in O_T. I corresponds to a set of instances and V to a set of data values. The functions i_X define the assignment of instances to the elements from O_T:

- $i_C : C_{KB} \rightarrow I$: instantiation of a concept
- $i_T : T_{KB} \rightarrow V$: instantiation of a data value
- $i_R : R_{KB} \rightarrow I \times I$: instantiation of a relation
- $i_A : A_{KB} \rightarrow I \times V$: instantiation of an attribute

> Ontologies describe structural relationships. Knowledge bases describe their concrete instantiation.

This means that the elements of a knowledge base (i.e., the concrete, factual knowledge in an information system) can now be represented as a directed graph $G = (N, E)$. The nodes $n_1, n_2 \in N$ are each connected by an edge $e \in E$ in the form of a triple (n_1, α, n_2), where α here denotes the specific label of the edge e. This formal definition is taken up semi-formally in the following concept of the Semantic Web and its associated Resource Description Framework and implemented in information technology [119].

The **Semantic Web** was designed to enable automated processing of resources described by ontologies. For this purpose, a layered technology stack is provided, as shown in Fig. 2.50. The Semantic Web is based on the established technologies also presented in this book and adds a number of additional components. The technology stack shows that, fundamentally, all communication is based on

Fig. 2.50 Technology stack of the Semantic Web [119]

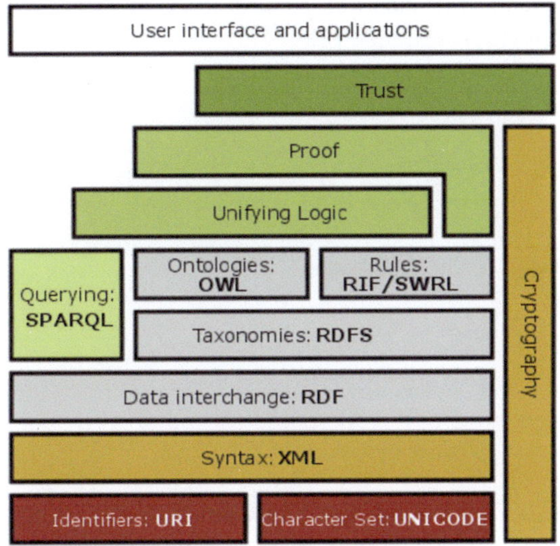

Unicode (see Subsection 2.4). Elements are addressed via unique identifiers (so-called URIs, which will be discussed in more detail later in Subsection 3.4). XML was used as the description language. Technologies for the joint representation of logic, truth, and trust, as well as for end-to-end encryption, were also considered. SPARQL was chosen as the query language (see Subsection 4.2). In the middle section are the already mentioned ontologies as well as taxonomies in the form of the Resource Description Framework Schema (RDFS) and data exchange at the instance level in the form of Resource Description Frameworks (RDF). We will take a closer look at these two layers in the following.

> The Semantic Web enables automated processing based on ontologies.

The **Resource Description Framework (RDF)** is also a W3C specification that defines a simple, graph-based data model for describing resources according to the knowledge base schema mentioned above [110, 119]. Resources in RDF are identified by a URL and can describe any object of interest. The basic elements are so-called RDF triples, which can be used to express a statement about a resource under consideration. Applied to the general triple form of a graph mentioned above (n_1, α, n_2), the definition of RDF triples often uses the notation $t := (s, p, o)$ to emphasize the linguistic properties of *S*ubject, *P*redicate, and *O*bject. Here, both the subject and the object are resources that are connected by the predicate. If the entire knowledge of a knowledge base is described using RDF triples, these can also be stored in databases. Systems that support this are therefore called triple-stores. Since clearly defined statements are available in the form of RDF triples, statements can also be inferred or checked (inferred), or new statements can be derived. RDF triples can be described in XML format (and several others); an example is shown in Listing 2.16. In addition, Listing 2.17 demonstrates a small SPARQL example [2].

```
1  <rdf:RDF>
2      <rdf:Statement rdf:about="mmfg:Statement">
3          <rdf:subject rdf:resource="Node:Hat"/>
4          <rdf:predicate rdf:resource="Relationship:above"/>
5          <rdf:object rdf:resource="Node:Head"/>
6      </rdf:Statement>
7      ...
8  </rdf:RDF>
```

Listing 2.16 A simple RDF triple

```
1      SELECT ?x ?y ?z
2      WHERE {
3          ?x rdfs:subClassOf:watch .
4          ?x rdfs:attribute:new .
5          ?y rdfs:subClassOf:Person .
6          ?y rdfs:name:Jim .
7          ?x rdfs:partof:?y .
8          ?z rdfs:type:Image
9      }
```

Listing 2.17 A SPARQL-Query based on RDF for "Jim's new watch"

RDF triples describe relationships in a knowledge base.

The **Resource Description Framework Schema (RDFS)** is another W3C specification that builds on the documentation of the RDF data model (cf. [W3Cf]). RDFS is a description language that provides a vocabulary for grouping and linking resources. In RDFS, so-called classes `rdfs:Class` and properties of these classes are introduced. Classes in RDFS are used to meaningfully group the resources described by RDF. Resources that belong to an `rdfs:Class` are called instances of that class. Instances of classes have certain properties that relate them to each other. One of the properties provided by RDFS is the subclass relationship, which states that a class C2 is a subclass of C1. This relationship is expressed as: `C2 rdfs:subClassOf C1`. In addition to the hierarchical order, further classification features can be expressed for the systematic categorization of resources. Properties in RDFS are restricted by specifying the so-called domain and range. The domain indicates that a property (R) is restricted to instances of certain classes and is expressed as `R rdfs:domain C1`. The range, in turn, specifies which instances this relation can be applied to `R rdfs:range C2`.

RDFS is limited in its expressiveness to the specification of classes and certain properties of relations. Other properties, such as the disjointness of classes, the combination of classes (union, intersection, complement of classes), cardinality restrictions, or certain characteristics of properties such as transitivity or uniqueness, cannot be expressed in RDFS. The W3C therefore specifies the **ontology language OWL.** The OWL specification exists in three variants: OWL Lite, OWL DL, and OWL Full. Starting from the language scope of OWL Lite, the following two variants, OWL DL and OWL Full, build on each other in terms of expressiveness in the order mentioned. The main advantages of OWL are that it allows for a clearer regulation of the possibilities of relationships. For example, restrictions (see Listing 2.18), value constraints, cardinality restrictions, set operations (union, intersection, etc.), or equivalence classes can be formulated, which of course enables significantly more semantic information, including structural information.

```
1  <owl:Class rdf:ID="Klasse">
2    <owl:equivalentClass>
3      <owl:Restriction>
4      <owl:onProperty rdf:resource="Property"/>
5          value constraint/cardinality constraint
6      </owl:Restriction>
7    </owl:equivalentClass>
8  </owl:class>
```

Listing 2.18 Mit OWL formulierte Restriktionen

RDFS describes structural properties of ontologies.

In this context, projects such as the Google Knowledge Graph [120] or Wikidata [112] should also be mentioned, which in turn enable OWL- and RDF-compatible queries on a knowledge base curated by the respective organization. What is particularly interesting here is that for each managed object, not only can all relationships to other objects be resolved, but also that each object has been assigned a unique identifier, which ensures worldwide, language-independent, taxonomically unique identities for semantic concepts. The jaguar mentioned several times above, for example, has the Wikidata ID `Q35694`, through which all metadata can be retrieved via common APIs or directly via URL: `https://www.wikidata.org/wiki/Q35694`. Unique identifiers will be discussed in more detail later in the book (see Subsection 3.4).

2.9   Summary

This chapter introduced the technical foundations and common methods for the extraction and representation of multimedia features. Depending on the respective multimedia types, there are a variety of different frameworks and concepts that can be used and applied. The role of text features is particularly noteworthy. On the one hand, because they historically laid the foundation for the field of information retrieval, and on the other hand, because the textual representation of features can be transferred to all types of features. This means that all the methods presented from the field of text features can also be directly applied to all other media types, provided their features are represented textually. In addition, a number of cross-cutting functions were introduced, such as Huffman coding, machine learning, feature graphs, or semantics, all of which contribute to the efficient management and understanding of multimedia. On this basis, the tasks and components of multimedia information systems can now be considered in the following chapters.

2.10   Self-Assessment Questions

2.1—Text Encoding

Using the US-ASCII character set (see Fig. 2.1), determine the hexadecimal encoding of the word "Multimedia".

2.2—Character Sets

What are glyphs?

2.3—Character Sets

What is the difference between US-ASCII and Unicode?

2.4—Text Formats

What is the advantage of XML documents?

2.5—Text Processing

Create a vectorization table for the following text: "Over there stands a tree. On the tree sits a bird. The bird is chirping loudly. At the base of the tree sits a cat and watches the bird."

2.6—Index

Create an inverted index at the sentence level for the terms from exercise 2.5.

2.7—Index

Determine the term frequency for the terms from exercise 2.5. Also consider the relation to the vectorization table you created.

2.8—Image Features

Explain the term anti-aliasing.

2.9—Image Features

What is the alpha channel?

2.10—Huffman Coding

Create a Huffman coding for the sentence "Abrakadabra Simsalabim".

2.11—Audio Features

Explain the terms sampling and quantization.

2.12—Audio Features

What does the sampling theorem state (brief explanation)?

2.13—Video Features

Calculate the file size of a video in HD (1280×720), Full HD (1920×1080), and Ultra HD (2840×2160), 60 Hz and a color depth of 24 bit with stereo sound (11 kHz, 8 bit) and a duration of 4:30 min.

2.14—Video Features

What is MPEG-7 used for?

2.15—Machine Learning

Which machine learning methods do you know?

2.16—Graphs

Create a weighted adjacency matrix for Fig. 2.45 assuming the following weights:

$$A \rightarrow B = 5$$
$$A \rightarrow C = 3$$
$$B \rightarrow D = 8$$
$$D \rightarrow C = 3$$
$$C \rightarrow E = 2$$
$$D \rightarrow E = 1$$

2.17—Semantics

What is a thesaurus?

2.18—Semantics

What role do functions play in ontologies?

References

1. A. M. Keller and F. Korn, "Query by example," *ACM Computing Surveys (CSUR)*, vol. 21, no. 1, pp. 29–58, 1989.
2. W3C. "SPARQL Query Language for RDF." [Online]. Available: https://www.w3.org/TR/sparql11-overview/, Download: 14.12.2021.
3. Internet Engineering Task Force (IETF). "The JavaScript Object Notation (JSON) Data Interchange Format." [Online]. Available: https://datatracker.ietf.org/doc/html/rfc7159, Download: 01.02.2021.
4. Open.AI, *Chat gpt*, https://chat.openai.com/, 2023.
5. Adobe Systems Software Ireland Ltd. "Creative journeys start here." [Online]. Available: http://www.adobe.com.
6. I. Maglogiannis, C. Doukas, G. Kormentzas, and T. Pliakas, "Wavelet-based compression with roi coding support for mobile access to dicom images over heterogeneous radio networks," *Information Technology in Biomedicine, IEEE Transactions on*, vol. 13, pp. 458–466, 2009. https://doi.org/10.1109/TITB.2008.903527.
7. R. Schenkel, W. Nejdl, and A. Dengel, *Multimedia Information Retrieval: Content-Based Information Retrieval from Large Text and Audio Databases*. Berlin, Heidelberg: Springer, 2007.
8. H. Balzert and C. Ebert, *Lehrbuch der Softwaretechnik. 3: Softwaremanagement/unter Mitw. von Christof Ebert* (Lehrbücher der Informatik), 2. Aufl. Heidelberg: Spektrum, Akad. Verl, 2008, 721 pp., ISBN: 978-3-8274-1161-7.
9. "ASCII format for network interchange," Internet Engineering Task Force, Request for Comments RFC 20, Oct. 1969, Num Pages: 9. https://doi.org/10.17487/RFC0020. Accessed: Feb. 21, 2023. [Online]. Available: https://datatracker.ietf.org/doc/rfc20.

10. "ASCII/ISO 8859 (latin-1) table," Accessed: Feb. 21, 2023. [Online]. Available: https://cs.stanford.edu/people/miles/iso8859.html.

11. Library of Congress (U.S.) "The most common text based formats." Publisher: South Carolina Electronic Records Archive. [Online]. Available: http://e-archives.sc.gov/formats-text/.

12. Library of Congress (U.S.) "Formats, evaluation factors, and relationships." Publisher: Library of Congress (U.S.) [Online]. Available: https://www.loc.gov/preservation/digital/formats/fdd/descriptions.shtml.

13. "UTR#17: Unicode character encoding model," Accessed: Feb. 15, 2023. [Online]. Available: https://unicode.org/reports/tr17/.

14. 14:00-17:00. "ISO/IEC 9075-2:2016," ISO, Accessed: Feb. 15, 2023. [Online]. Available: https://www.iso.org/standard/63556.html.

15. Adobe Inc. "Portable document format specification," Adobe Systems Incorporated. [Online]. Available: https://www.adobe.com/content/dam/acom/en/devnet/pdf/pdfs/PDF32000_2008.pdf, Download: 11.07.2021.

16. Microsoft Corp. "Word Extensions to the Office Open XML File Format," Microsoft Corporation. [Online]. Available: https://interoperability.blob.core.windows.net/files/MS-DOCX/%5bMS-DOCX%5d.pdf, Download: 11.07.2021.

17. W3C. "Hypertext Markup Language," W3.org. [Online]. Available: https://www.w3.org/TR/html52/, Download: 11.07.2021.

18. W3C. "eXtensible markup language." Publisher: W3.org. [Online]. Available: https://www.w3.org/TR/xml/.

19. S. Büttcher, C. L. A. Clarke, and G. V. Cormack, *Information Retrieval: Implementing and Evaluating Search Engines*. Cambridge, Massachusetts: The MIT Press, 2016.

20. K. Aberer, *The Semantic Web*. Berlin, Heidelberg, New York: Springer Verlag, 2007, ISBN: 978-3-540-76297-3.

21. J. Smith and J. Johnson, "Text vectorization: Methods, techniques, and challenges," *Journal of Natural Language Processing*, vol. 25, no. 2, pp. 145–165, 2021.

22. J. Doe and M. Smith, "A survey of text tokenization techniques," in *Proceedings of the International Conference on Natural Language Processing*, ACM, 2022, pp. 45–56.

23. J. Doe and J. Johnson, "N-gram based text representation for information retrieval," in *Proceedings of the ACM International Conference on Information Retrieval*, ACM, 2021, pp. 123–135.

24. M. F. Porter, "An algorithm for suffix stripping," *Program*, vol. 14, no. 3, pp. 130–137, 1980.

25. J. Smith and M. Johnson, "Lemmatization strategies for natural language processing," *Journal of Computational Linguistics*, vol. 30, no. 4, pp. 589–612, 2018.

26. J. Silge and D. Robinson. "Text Mining with R: A Tidy Approach." [Online]. Available: https://www.tidytextmining.com/tfidf.html, Download: 14.11.2021.

27. C. D. Manning, P. Raghavan, and H. Schütze, *Introduction to information retrieval*. New York: Cambridge University Press, 2008, 482 pp., OCLC: ocn190786122, ISBN: 978-0-521-86571-5.

28. C. D. Manning, P. Raghavan, and H. Schütze, *Introduction to Information Retrieval*. Cambridge University Press, 2008.

29. H. Balzert and P. Liggesmeyer, *Lehrbuch der Softwaretechnik. 2: Entwurf, Implementierung, Installation und Betrieb/Helmut Balzert. Unter Mitw. von Peter Liggesmeyer* (Lehrbücher der Informatik), 3. Auflage. Heidelberg: Spektrum, Akademischer Verlag, 2011, 596 pp., ISBN: 978-3-8274-1706-0.

30. H. Balzert, *Lehrbuch der Softwaretechnik. 1: Basiskonzepte und Requirements-Engineering/Helmut Balzert. Unter Mitw. von Heide Balzert* (Lehrbücher der Informatik), 3. Aufl, H. Balzert, Ed. Heidelberg: Spektrum Akad. Verl, 2009, 624 pp., ISBN: 978-3-8274-1705-3.

31. M. T. Goodrich and R. Tamassia, *Data Structures and Algorithms in Java*. Wiley, 2014.

32. M. Haindl, S. Mikeš, and G. Scarpa, "Unsupervised Detection of Mammogram Regions of Interest," in *Knowledge-Based Intelligent Information and Engineering Systems*, Berlin, Heidelberg: Springer Berlin Heidelberg, 2007, pp. 33–40.

33. Adobe Inc. "Adobe stock." [Online]. Available: https://stock.adobe.com, Download: 02.10.2020.

34. S. Wagenpfeil, B. Vu, P. Mc Kevitt, and M. Hemmje, "Fast and effective retrieval for large multimedia collections," *Big Data and Cognitive Computing*, vol. 5, no. 3, 2021, ISSN: 2504-2289. https://doi.org/10.3390/bdcc5030033. [Online]. Available: https://www.mdpi.com/2504-2289/5/3/33.

35. S. Wagenpfeil, P. M. Kevitt, and M. Hemmje, "Smart multimedia information retrieval," *Analytics*, vol. 2, no. 1, pp. 198–224, Feb. 20, 2023, ISSN: 2813-2203. https://doi.org/10.3390/analytics2010011. Accessed: Feb. 21, 2023. [Online]. Available: https://www.mdpi.com/2813-2203/2/1/11.

36. R. C. Gonzalez, R. E. Woods, and S. L. Eddins, *Digital Image Processing*, 5th Edition. Harlow, England: Pearson, 2021.

37. Library of Congress (U.S.) "The most common image based formats." Publisher: South Carolina Electronic Records Archive. [Online]. Available: http://e-archives.sc.gov/formats-image/.

38. Wikipedia. "Tag image file format," Wikipedia. [Online]. Available: https://en.wikipedia.org/wiki/TIFF, Download: 11.07.2021.

39. LibPNG.org. "PNG (Portable Network Graphics) Specification, Version 1.2," LibPNG.org. [Online]. Available: http://www.libpng.org/pub/png/spec/1.2/PNG-Structure.html, Download: 11.07.2021.

40. W3C. "Graphics interchange format," W3C.org. [Online]. Available: https://www.w3.org/Graphics/GIF/spec-gif87.txt, Download: 11.07.2021.

41. JPEG.org. "Overview of jpeg," JPEG.org. [Online]. Available: https://jpeg.org/jpeg/, Download: 11.07.2021.

42. Adobe.com. "Photoshop file formats." [Online]. Available: https://helpx.adobe.com/photoshop/using/file-formats.html, Download: 03.10.2020.

43. Fujifilm Corp. "X-T3," Fujifilm Corporation. [Online]. Available: https://fujifilm-x.com/global/products/cameras/x-t3/, Download: 03.07.2021.

44. J. D. Foley, A. van Dam, S. K. Feiner, and J. F. Hughes, *ComputerGraphics: Principles and Practice*. Addison-Wesley Professional, 1995.

45. M. Rabbani and P. W. Jones, *Digital Image Compression Techniques*. SPIE Press, 2002.

46. W. B. Pennebaker and J. L. Mitchell, *JPEG: Still Image Data Compression Standard*. Van Nostrand Reinhold, 1992.

47. J. Smith and M. Johnson, "Run-length encoding for image compression and transmission," *Journal of Image Processing*, vol. 10, no. 3, pp. 215–227, 2021.

48. J. Ziv and A. Lempel, "A universal algorithm for sequential data compression," *IEEE Transactions on Information Theory*, vol. 23, no. 3, pp. 337–343, 1977.

49. D. Huffman, "A method for the construction of minimum-redundancy codes," *Proceedings of the IRE*, vol. 40, no. 9, pp. 1098–1101, Sep. 1952, ISSN: 0096-8390. https://doi.org/10.1109/JRPROC.1952.273898. Accessed: Feb. 15, 2023. [Online]. Available: http://ieeexplore.ieee.org/document/4051119/.

50. W. Dankmeier, *Grundkurs Codierung: Verschl üsselung, Kompression, Fehlerbeseitigung*, 4., überarbeitete Auflage 2017. Wiesbaden: Vieweg & Teubner, 2017, 451 pp., ISBN: 978-3-8348-1674-0.

51. Microsoft Corp. "Microsoft word." Publisher: Microsoft Corporation. [Online]. Available: https://www.microsoft.com/en-us/microsoft-365/word.

52. A. Holzinger, *Basiswissen Multimedia. 1: Technik: technologische Grundlagen multimedialer Informationssysteme*, 2., überarb. u. erw. Aufl. Würzburg: Vogel, 2002, 320 pp., ISBN: 978-3-8023-1914-3.

53. P. A. Henning, *Taschenbuch Multimedia*, 3., bearb. Aufl. München: Fachbuchverl. Leipzig im Carl Hanser Verl, 2003, 646 pp., ISBN: 978-3-446-22308-0.

54. C. Wu and M. E. Thompson, *Sampling theory and practice* (ICSA book series in statistics). Cham, Switzerland: Springer, 2020, 365 pp., ISBN: 978-3-030-44392-4 978-3-030-44244-6.

55. M. Meyer, *Signalverarbeitung: analoge und digitale Signale, Systeme und Filter: mit 161 Abbildungen und Tabellen* (Lehrbuch), 8., verbesserte Auflage. Wiesbaden: Springer Vieweg, 2017, 324 pp., ISBN: 978-3-658-18320-2.

56. J. Lange and T. Lange, *Mathematische Grundlagen der Digitalisierung: kompakt, visuell, intuitiv verst ändlich* (essentials). Wiesbaden [Heidelberg]: Springer Vieweg, 2019, 59 pp., ISBN: 978-3-658-26685-1.

57. A. Spanias, T. Tseng, and T. Painter, *Audio Signal Processing and Coding*. Wiley-Interscience, 2007.

58. MPEG. "MPEG-4." Publisher: The Moving Pictures Experts Group. [Online]. Available: https://mpeg.chiariglione.org/standards/mpeg-4/mp4-file-format.

59. T. Zielinski, "Fast fourier transform," in Jan. 2021, pp. 93–114, ISBN: 978-3-030-49255-7. https://doi.org/10.1007/978-3-030-49256-4_5.

60. R. N. Bracewell, *Fourier and Cosine Transformations: A Collection of Tables*. New York, NY: Springer, 1986.

61. R. Steinmetz, *Multimedia-Technologie*. Berlin, Heidelberg: Springer Berlin Heidelberg, 1999, ISBN: 978-3-662-08882-1 978-3-662-08881-4. https://doi.org/10.1007/978-3-662-08881-4. Accessed: Feb. 15, 2023. [Online]. Available: http://link.springer.com/10.1007/978-3-662-08881-4.

62. Library of Congress (U.S.) "The most common audio based formats." Publisher: South Carolina Electronic Records Archive. [Online]. Available: http://e-archives.sc.gov/formats-audio/.

63. "Logic Pro," Apple (Deutschland), Accessed: Feb. 27, 2023. [Online]. Available: https://www.apple.com/de/logic-pro/.

64. Fraunhofer Institut. "MP3 File Format Specifications," Fraunhofer IIS. [Online]. Available: https://www.iis.fraunhofer.de/en/ff/amm/consumer-electronics/mp3.html,   Download: 11.07.2021.

65. Wikipedia. "Video codec," Wikipedia. [Online]. Available: https://en.wikipedia.org/wiki/Video_codec, Download: 11.07.2021.

66. M. Ghanbari, *Video Codec Design: Developing Image and Video Compression Systems*. Chichester, England: John Wiley & Sons, 1999.

67. K. Jack, *Video Demystified: A Handbook for the Digital Engineer*. Oxford, UK: Newnes, 2011.

68. Library of Congress (U.S.) "The most common video based formats." Publisher: South Carolina Electronic Records Archive. [Online]. Available: http://e-archives.sc.gov/formats-video/.

69. T. Wiegand, G. J. Sullivan, G. Bjøntegaard, and A. Luthra, "Entropy coding in image and video compression: A tutorial," *Proceedings of the IEEE*, vol. 93, no. 1, pp. 184–197, 2005.

70. H. Higgins and P. Gray, "Differential pulse code modulation: A survey," *Proceedings of the IEEE*, vol. 60, no. 11, pp. 1385–1417, 1972.

71. The Moving Picture Experts Group. "MPEG2 - Generic coding of moving pictures and associated audio." [Online]. Available: https://mpeg.chiariglione.org/standards/mpeg-2, Download: 12.06.2021.

72. B. G. Sherlock, "Transform coding of images," *Proceedings of the IEEE*, vol. 73, no. 4, pp. 523–548, 1985.

73. W3C.org. "MPEG-4: a Powerful Standard for Use in Web and Television Environments." [Online]. Available: https://www.w3.org/Architecture/1998/06/Workshop/paper26/, Download: 14.06.2021.

74. Shih-Fu Chang, T. Sikora, and A. Purl, "Overview of the MPEG-7 standard," *IEEE Transactions on Circuits and Systems for Video Technology*, vol. 11, no. 6, pp. 688–695, 2001.

75. S. Büttcher, C. Clarke, and G. Cormack, *Information Retrieval*. Massachusetts: MIT Press, 2010, ISBN: 978-0-262-02651-2.

76. B. U. M. D. Group. "Bilvideo-7: An mpeg-7 compatible video indexing and retrieval system." [Online]. Available: http://www.cs.bilkent.edu.tr/ bilmdg/bilvideo-7/.

77. W. g. US Government. "Inston churchill and franklin d. roosevelt at yalta, 1945." [Online]. Available: https://en.wikipedia.org/w/index.php?title=File:Churchill_and_Roosevelt_Yalta. jpg.

78. Wikipedia. "Advanced video coding," Wikipedia. [Online]. Available: https://en.wikipedia. org/wiki/Advanced_Video_Coding, Download: 11.07.2021.

79. Microsoft Corp. "AVI RIFF File Reference," Microsoft Corporation. [Online]. Available: https://docs.microsoft.com/en-us/windows/win32/directshow/avi-riff-file-reference, Download: 11.07.2021.

80. Apple Inc. "Introduction to QuickTime File Format Specification," Apple Inc. [Online]. Available: https://developer.apple.com/library/archive/documentation/QuickTime/QTFF/ QTFFPreface/qtffPreface.html, Download: 11.07.2021.

81. A. S. Foundation. "Apache poi – the java api for microsoft documents. "[Online]. Available: https://poi.apache.org, Download: 13.10.2020.

82. S. Wagenpfeil, *Moderne Software-Entwicklung mit Java und der JEE*, 1st Edition. Springer, 2022, ISBN: 978-3662665817.

83. MIT – Massachutsetts Institute of Technology. "Description of Exif file format." [Online]. Available: http://media.mit.edu/pia/Research/deepview/exif.html, Download: 27.09.2020.

84. J. Beyerer, M. Richter, and M. Nagel, *Pattern Recognition – Introduction, Features, Classifiers and Principles*. Berlin: Walter de Gruyter GmbH & Co KG, 2017, ISBN: 978-3-110-53794-9.

85. R. Scherer, *Computer Vision Methods for Fast Image Classification and Retrieval*. Gewerbestrasse 11 – 6330 Cham CH: Springer Nature, 2020, pp. 23–31, ISBN: 978-3-030-12194-5.

86. N. Gayathri and K. Mahesh, "An efficient video indexing and retrieval algorithm using ensemble classifier," in *2019 4th International Conference on Electrical, Electronics, Communication, Computer Technologies and Optimization Techniques (ICEECCOT)*, 2019, pp. 250–258.

87. C. M. Bishop, *Pattern Recognition and Machine Learning*. Springer, 2006.

88. Microsoft Inc. "Machine Visioning." [Online]. Available: http://azure.microsoft.com/ser-vices/cognitive-services/computer-vision, Download: 10.03.2021.

89. Amazon.com, "Amazon recognition," Amazon.com, http://aws.amazon.com/recognition, Tech. Rep., Jul. 2020.

90. J. Redmon, S. Divvala, R. Girshick, and A. Farhadi, "You only look once: Unified, real-time object detection," Jun. 2016, pp. 779–788. https://doi.org/10.1109/CVPR.2016.91.

91. Image AI. "Image AI – State-of-the-art Recognition and Detection AI with few lines of code.," ImageAI.org. [Online]. Available: http://www.imageai.org, Download: 11.08.2021.

92. Clarifai. "Clarifai – endless possibilities with computer vision, natural language processing and automated machine learning.," Clarifai Inc. [Online]. Available: https://www.clarifai. com, Download: 11.08.2021.

93. Wikipedia. "Tensorflow," Wikipedia. [Online]. Available: https://en.wikipedia.org/wiki/ TensorFlow, Download: 11.08.2021.

94. FFMpeg.org. "Ffmpeg documentation." Place: http://ffmpeg.org Publisher: FFMpeg.org.

95. E. Union. "Eu artificial intelligence act." [Online]. Available: https://artificialintelligenceact. eu/de/article/3/, Download: 03.10.2024.

96. S. der Wissenschaft. "Künstliche intelligenz." [Online]. Available: https://www.spektrum. de/lexikon/neurowissenschaft/kuenstliche-intelligenz/6810, Download: 03.10.2024.

97. I. Art, *Midjourney*, https://www.imagine.art/, 2023.

98. Open.AI, *Dall-e*, https://openai.com/dall-e-2, 2023.

99. OpenAI, *DALL·E: Creating Images from Text*, https://openai.com/dall-e, Zugriff am 9. Juni 2025, 2025. [Online]. Available: https://openai.com/dall-e.

100. M. Needham, *Graph Algorithms*. 1005 Gravenstein Highway North Sebastopol CA 95472: O'Reilly Media Inc., 2019, ISBN: 978-1-492-05781-9.

101. W3C. "eXtensible Markup Language," W3.org. [Online]. Available: https://www.w3.org/ TR/xml/, Download: 11.07.2021.

102. E. E. Ian Robbinson Jim Webber, *Graph Databases*. 1005 Gravenstein Highway North Sabstopol CA 95472: O'Reilly Media Inc., 2015, ISBN: 978-1-491-93089-2.

103. Neo4J Inc. "Neo4j – graph database." [Online]. Available: https://neo4j.com, Download: 01.10.2020.

104. The GraphQL Foundation. "Graph QL – a query language for your API." [Online]. Available: https://graphql.org, Download: 24.10.2020.

105. W. Hamilton, R. Ying, and J. Leskovec, "Representation learning on graphs: Methods and applications," *Dpt. of Computer Science, Stanford University*, Sep. 2017. arXiv:1709.05584v3.

106. J. Qiu, Y. Dong, H. Ma, J. Li, K. Wang, and J. Tang, "Network embedding as matrix factorization: Unifying deepwalk, line, pte, and node2vec," *CoRR*, vol. abs/1710.02971, 2017. arXiv: 1710.02971. [Online]. Available: http://arxiv.org/abs/1710.02971.

107. P. Goyal and E. Ferrara, "Graph embedding techniques, applications, and performance: A survey," *CoRR*, vol. abs/1705.02801, 2017. arXiv: 1705.02801. [Online]. Available: http://arxiv.org/abs/1705.02801.

108. Sciencedirect.com. "Adjacency matrix." [Online]. Available: https://www.sciencedirect.com/topics/mathematics/adjacency-matrix.

109. Duden, *Semantik*, https://www.duden.de/suchen/dudenonline/semantik/, 2023.

110. SemanticWeb.com. "From taxonomies over ontologies to knowledge graphs." [Online]. Available: https://semantic-web.com/2014/07/15/from-taxonomies-over-ontologies-to-knowledge-graphs/, Download: 02.10.2020.

111. I. S. Organisation. "Iso 25964-1:2011," Information, documentation – Thesauri, and interoperability with other vocabularies – Part 1: Thesauri for information retrieval. [Online]. Available: https://www.iso.org/standard/53657.html, Download: 11.08.2021.

112. Wikidata.com. "Wikidata – the free knowledgebase." [Online]. Available: https://www.wikidata.org/wiki/Wikidata:Main_Page.

113. Food and Agriculture Organization of the United Nations, *AGROVOC Multilingual Thesaurus*, https://www.fao.org/agrovoc/, Zugriff am 9. Juni 2025, 2025. [Online]. Available: https://www.fao.org/agrovoc/.

114. W3C. "W3C Semantic Web Activity." [Online]. Available: http://w3.org/2001/sw, Download: 23.11.2021.

115. M. N. Asim, M. Wasim, M. U. G. Khan, N. Mahmood, and W. Mahmood, "The use of ontology in retrieval: A study on textual, multilingual, and multimedia retrieval," *IEEE Access*, vol. 7, pp. 21 662–21 686, 2019.

116. T. R. Gruber, "A translation approach to portable ontology specifications," *Knowledge Acquisition*, vol. 5, no. 2, pp. 199–220, Jun. 1993, ISSN: 10428143. https://doi.org/10.1006/knac.1993.1008. Accessed: Feb. 16, 2023. [Online]. Available: https://linkinghub.elsevier.com/retrieve/pii/S1042814383710083.

117. A. G. Pérez and R. Benjamins, "Overview of knowledge sharing and reuse components: Ontologies and problem-solving methods," in *International Joint Conference on Artificial Intelligence*, 1999.

118. M. Ehrig, P. Haase, N. Stojanovic, and M. Hefke, "Similarity for ontologies – a comprehensive framework," in *Information Systems in a Rapidly Changing Economy: Proceedings of the 13th European Conference on Information Systems (ECIS 2005), May 26–28, 2005, Regensburg, Germany*, D. Bartman et al., Eds., Association for Information Systems, 2005. [Online]. Available: http://is2.lse.ac.uk/asp/aspecis/20050131.pdf.

119. J. Domingue, D. Fensel, and J. Hendler, *Introduction to the Semantic Web Technologies*. Berlin, Heidelberg: Springer Berlin Heidelberg, 2011, pp. 1–41, ISBN: 978-3-540-92913-0. https://doi.org/10.1007/978-3-540-92913-0. [Online]. Available: https://doi.org/10.1007/978-3-540-92913-0.

120. Google Ireland Ltd. "Google knowledge search api." [Online]. Available: http://developers.google.com/knowledge-graph, Download: 14.04.2021.

Information Retrieval 3

Information Retrieval is a field of computer science that can be classified within the domain of information systems. The original focus of Information Retrieval (IR) was on the indexing and access of text documents, but in recent years it has increasingly been extended to multimedia documents such as images, videos, or audio recordings. Today, there is lively debate as to whether Multimedia Information Retrieval (MMIR) is a specialization of classical IR—since proven methods for text analysis are now being applied to arbitrary multimedia objects— or whether classical IR has always merely been a part of MMIR, since, after all, texts themselves are multimedia objects. As this book cannot resolve such a discussion and it is irrelevant for the further course, it is left to the reader to form their own opinion on this matter. The concepts of information and information need, as well as the various possibilities for feature extraction and representation, have already been discussed in detail in previous chapters, so the focus now shifts to the concrete application of these concepts and mechanisms within Information Retrieval. Information Retrieval, in direct German translation, means "information acquisition" or "finding information." Baeza-Yates [5] defines IR as "part of computer science which studies the retrieval of information (not data) from a collection of written documents. The retrieved documents aim at satisfying a users information need." Thus, IR describes a process of automated support in the search for information. As with any process, a typical sequence of processing steps can be identified in IR, which will be examined in this chapter.

3.1 Tasks of an IR System

The models of information seeking introduced at the beginning in Subsection 1.4 describe the trigger for the need for IR as well as the distinguishable requirements and contextual conditions within it. The remainder of this subsection will focus

© The Author(s), under exclusive license to Springer-Verlag GmbH, DE, part of Springer Nature 2026
M. Hemmje and S. Wagenpfeil, *Multimedia Information Retrieval*,
https://doi.org/10.1007/978-3-662-73310-3_3

primarily on the methods that implement what Bates (see Fig. 1.10) defines as searching, i.e., supporting users in their active effort to find answers to open questions or to gain an understanding of a new subject matter.

In the literature, the term **ad hoc retrieval** has become established for the implementation of search. Ad hoc retrieval is treated as one of the overarching tasks (English: "retrieval task") in IR. Manning [3] describes the nature of a retrieval task as: "the task executed by the information system in response to a user request. It is basically of two types: ad hoc and filtering." Thus, in a retrieval task, the two tasks of ad hoc retrieval and **filtering** are generally distinguished:

- **Ad hoc retrieval** is the standard task of IR: Users of an IR process specify an information need by means of a search query, which initiates an automated procedure. In a classic IR application, the output of this procedure is a set of relevant documents with textual content. The term ad hoc derives from the fact that the number of possible questions is virtually unlimited.
- **Filtering** describes the task in which the information need is relatively static, but the document collection is constantly expanding (e.g., searching through email lists).

This subsection deals exclusively with topics that can be assigned to the search process according to Bates or to ad hoc retrieval.

The original task of Information Retrieval is the search for full-text documents. In the field of Multimedia Information Retrieval, this also includes all other types of multimedia, or the textual representation of their features. Thus, the principle of ad hoc retrieval is also applicable to the processing of multimedia content, whether videos, audio files, scientific analysis data, or other types. However, the fundamentals of IR are best introduced through IR processes that operate on large document collections with textual content. Historically, many of the fundamental IR solutions have been developed on this basis. Therefore, in the course of this chapter, textual documents in a document collection will initially form the basis for further explanations.

> Ad hoc retrieval and filtering are the core tasks of MMIR systems.

In addition to this still very general description of search or ad hoc retrieval in a document collection, there are a multitude of specific application areas for their use. Due to the large number of different application areas, only a selection will be introduced here. One of the most important and established IR evaluation conferences, TREC (Text REtrieval Conference) [4], serves this purpose. There, investigations are conducted based on certain typical IR tasks, which thus also provides a typical selection of possible use cases:

- Complex Answer Retrieval Track: Development of information systems that enable the resolution of complex information needs by aggregating relevant information from a complete corpus;
- Conversational Assistance Track: Development and testing of information systems for automated dialogues;
- Decision Track: Development and testing of decision support systems;
- Incident Streams Track: Development and testing of information systems that process social media data for emergency situations;
- News Track: Development and testing of information systems that operate on news reports;
- Precision Medicine Track: Development and testing of information systems that establish connections between clinical trials and evidence-based literature to find new effective treatments.

Another subdivision in IR concerns the way in which retrieval is implemented and the requirements set by users. Two classes of IR models are distinguished here. The approaches of so-called **exact match models** assume that users have precise knowledge of the content and structure of a document collection. In this case, users can satisfy their information need using a well-defined query language. In direct contrast are the approaches assigned to the so-called **best match models**. In this class, it is assumed that users have neither complete knowledge of the structure and content of the data collection nor a fully differentiated idea of how to express their information need. This book primarily deals with solutions for best match models. Solutions for exact match requirements are introduced but not discussed in depth.

The excerpt of possible tasks that can be automated by an information system shows that the requirements for information system application solutions are very broad. It is not sufficient to select documents using a single method. Rather, the information system must provide application solutions that can prepare the data collection for various tasks, support the range of possible output formats, and enable interaction with users. Due to the multitude of requirements, the next subsection will first introduce a very general conceptual IR model.

3.2   General IR Model

Generally speaking, ad hoc retrieval can be understood as a process. This process involves a specific sequence of processing steps on a preprocessed set of textual documents. A general model that represents this process in its entirety was introduced by Fuhr [5] and will be documented here as representative of other models (see also Fig. 3.1).

First, it is established that a document is denoted by the variable d_n and $d_n \in D$ applies. The variable D here stands for the collection of all documents. A query

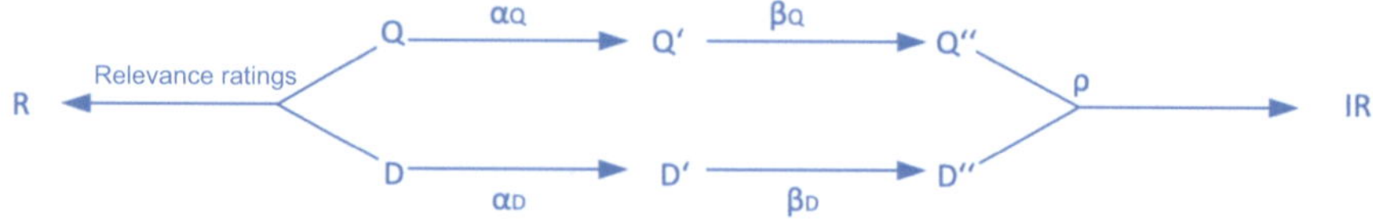

Fig. 3.1 Conceptual IR model according to Fuhr [5]

is denoted by q_k and $q_k \in Q$ applies, where Q is the set of all possible queries. It is further established that there is a relevance relationship between d_n and q_k, e.g., $R = \{R^+, R^-\}$, where R^+ indicates that a document is relevant to the query, and R^- that the document is not relevant to the query. This specification in particular implies the validity of the mapping $r : Q \times D \to R$.

However, an IR process has only limited means to represent a document or a query. Therefore, when processing d and q, an abstracted representation $d'_n \in D' \subset D$ and $q'_k \in Q' \subset Q$ is assumed, which is generated by the mappings $\alpha_D : d \to d'$ and $\alpha_Q : q \to q'$. An abstracted representation could, for example, be a set of words from the document and the query under consideration—similar to the index of a physical book, which also represents the content via included keywords. Thus, the content of a complete document would be contained in d_n, while the index of the same document, via the mapping α_D, would be in the subset d'_n. Fuhr's model goes one step further in abstraction, as there are IR models that also operate at an even higher level of abstraction for documents and queries. This further abstraction of the representation to $d''_n \in D''$ and $q''_k \in Q''$ is performed via the mappings $\beta_D : d' \to d''$ and $\beta_Q : q' \to q''$. These can be thought of as a kind of normalization function that, for example, aligns the values of a document's index and the query criteria entered, in order to make comparison possible in the first place. A so-called retrieval function then computes the relevance between the document and query representations: $p(q''_k, d''_n)$. The model underlying the retrieval function is called the **retrieval model**. The result of this computation is a value that expresses the relevance relationship and is commonly referred to as the **Retrieval Status Value (RSV)** [1].

> The Retrieval Status Value (RSV) denotes the relevance of a result.

The goal of an IR process is to provide automated support for information seeking. Systems that automate an IR process as shown in Fig. 3.1 are called Information Retrieval Systems (IRS) and are a specific type of information system (IS). First, the following relationship is recapitulated from the definitions in Subsection 1.2: Among other things, in economics, data is the basis for deriving information. Derived information manifests as knowledge. In economics, a frequently referenced model describing these relationships is the so-called

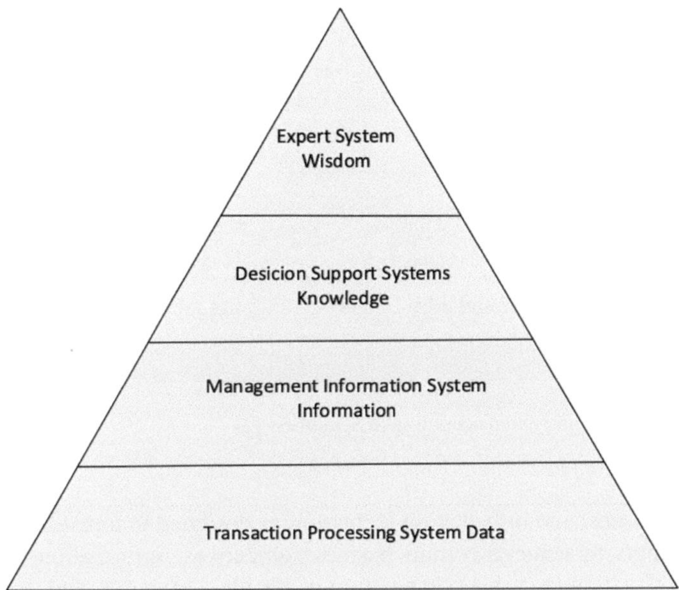

Fig. 3.2 The Data–Information–Knowledge–Wisdom Hierarchy [6]

Data–Information–Knowledge–Wisdom Hierarchy (DIKW Hierarchy) [6]. It is evident from this model, as already introduced, that information is based on access to data. Data is understood as values that are unprocessed and discrete (see Fig. 3.2).

The structure of the data accessible via an IS is described by a data model. Kemper et al. [7] define it as follows: "The data model defines the modeling constructs by which a computerized representation of the real world (or the relevant part thereof) can be generated […], it defines the generic structures and operators [!] that can be used to model a specific application."

An IS can thus be described as a system that, based on data specified by a data model, provides information for user access (cf. Fig. 3.3). Analogously, the following description also applies: "An information system is a formalized computer information system that can collect, store, process, and report data from various sources to provide the information necessary for managerial decision making" [1].

Other definitions of information systems particularly emphasize that such a system is not isolated, but has connections to other elements in its environment. An information system is often described as a collection of people, information,

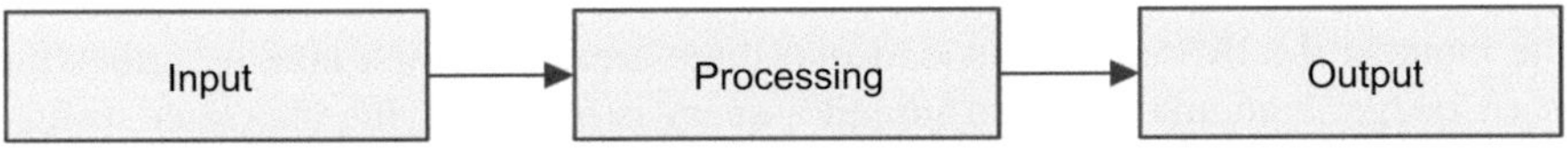

Fig. 3.3 A simple information system model

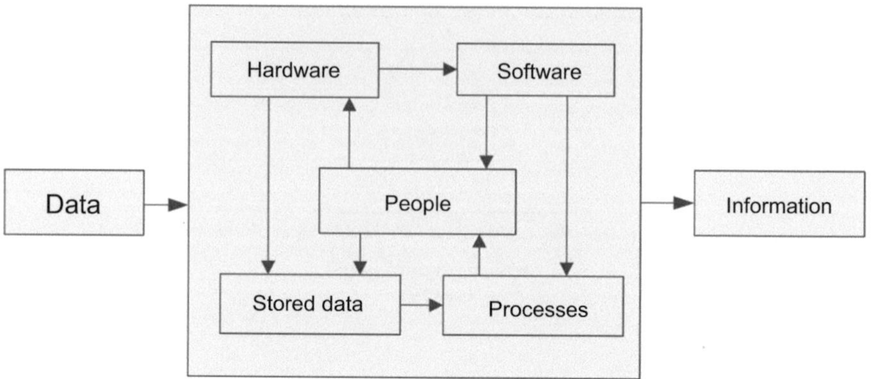

Fig. 3.4 An information system according to Schultheis [2]

business processes, and information technologies designed to transform inputs into specific outputs to achieve certain business objectives. Information systems are not seen in isolation, but as a composition of people, structures, and technologies. Each process from Fig. 3.3 can be mapped as a set of interconnected subsystems, which can be further broken down depending on the level of detail (see Fig. 3.4).

An **information retrieval system** is thus a specific type of IS for automated searching in document collections, provided that an information need exists. The IS models introduced so far have taken a rather business-oriented perspective on an IS. As already described in Subsection 1.4, this book adopts Kuhlen's perspective on the relationships between data, knowledge, and information. Kuhlen, whose semiotic view, in contrast to the business perspective, sees data as a foundation for both knowledge and information, considers the automated transformation of knowledge into information in the subjective context of users. Here, contextual conditions and the current knowledge state of users drive the use of an IRS. The following conceptual view of an IRS is taken from the publication by Buckland [8] (see Fig. 3.5).

The conceptual model generally provides a subdivision as in Fig. 3.3. There is an input to the system, the system contains a set of processing components, and these produce an output. Boxes in the figure without dashed borders describe general system functionalities. The meaning of the individual components will be explained in the following subsections in the order named.

3.2.1   System Input

The input to the IRS can be divided into three areas. The first area describes the user's query. A common way to submit a query is by specifying character strings, which may be part of a formal language (e.g., SQL or SPARQL) or in natural

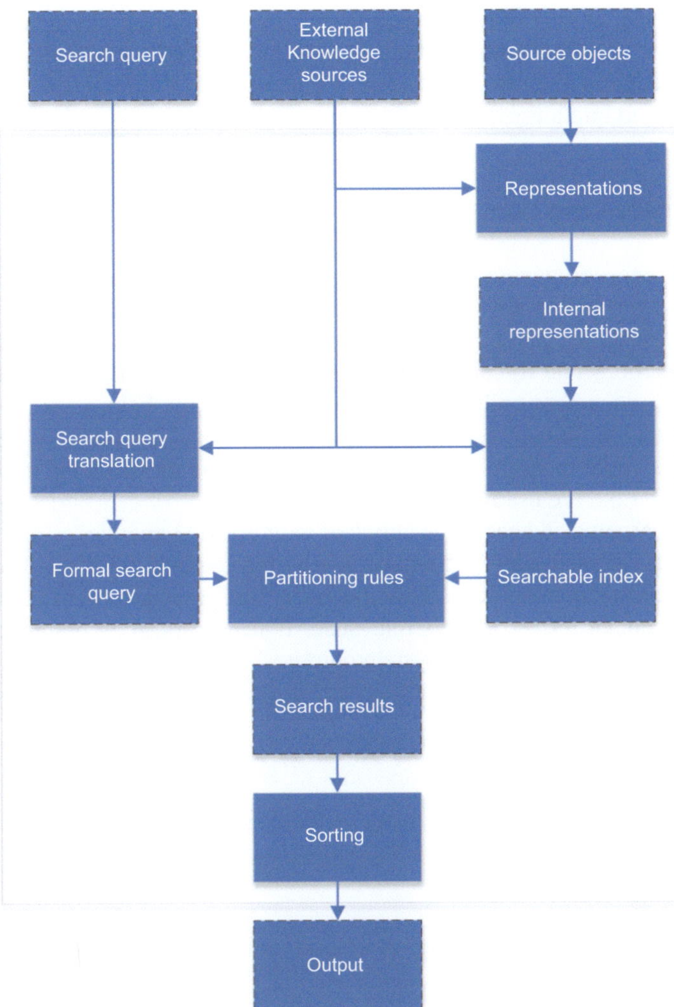

Fig. 3.5 Conceptual IRS model according to Buckland [8]

language. In the case of music retrieval, for example, this could also be melodies that are hummed, or an image in the case of searching for similar images ("query by example"). The IRS must be able to access a document collection against which it can meaningfully map the query. Potentially, the elements of the collection could have any multimedia content.

In addition to the query and the elements to be searched, the inclusion of external knowledge sources is also mentioned. These external knowledge sources can be specific databases or human expertise. The knowledge available from these sources is used, among other things, to optimize queries (e.g., by determining the meaning of a word in a colloquial query, such as "Jaguar"—car brand or animal?). This knowledge is also used to convert the plain text of a document in the document collection into a form usable for retrieval. This process is called **indexing**.

3.2.2 Processing Components

The processing component receives the input and processes it for various purposes. However, before a query can be processed at all, all elements of the document collection must first be converted into a representation understandable by the retrieval function. The so-called indexing process examines the content of the elements in the document collection and, possibly with the help of external knowledge, converts it into a corresponding representation—the **index**. The index is stored in a storage medium accessible to the IRS. The mechanisms introduced in Subsection 2.1.5.5 are used to build the index.

The query may also need to be enriched with external knowledge and then converted into a representation understandable by the retrieval function. The retrieval function receives representations of the query and the document elements and compares them in terms of their relevance. If necessary, the computed result is then sorted and passed on to the system output.

3.2.3 System Output

The system output visualizes the result of the processing component for the users and, if necessary, enriches it with additional information.

> IR systems can be divided into three core components: system input, processing, and system output.

In this subsection, you have gained insight into the very general tasks of Information Retrieval. Basically, every IR or MMIR system can be reduced to the rudimentary process input → processing → output. Various definitions have been provided to formalize this process and thus make it describable and modelable in IT terms. At its core, however, it is this "simple" task that we expect from MMIR systems.

The classification presented here again shows that virtually every IT process can also be mapped via MMIR systems, since the very general division into system input, processing, and system output fundamentally forms the basis for every interactive system. Thus, the saying mentioned at the beginning, "Multimedia is everywhere," can also be applied to MMIR systems: "MMIR is everywhere!"

3.3   Other Models and Concepts of Information Retrieval

From a general perspective, models in science are abstract representations of phenomena. Examples include explanations of conceptual constructs, mathematical theses, or other relationships, such as climate or the development of currency values. A model in science is used, among other things, to communicate or explain these phenomena or to test them with respect to a particular hypothesis.

Models are also used in IR for a wide variety of requirements. At the beginning of this book, various models for describing information, knowledge, or interactions have already been introduced. IR models in particular, however, describe the relationships between the input of the information need to the Information Retrieval System (IRS), how relevant documents are selected, and how they are output. If the models are implemented in software, they can also be evaluated and assessed. Such an evaluation of IR models can be carried out at various levels.

Since a wide variety of IR models have emerged over time, this chapter will first distinguish the models based on differentiable criteria. Based on these characteristics, the models can then be introduced in an organized manner. In principle, different criteria can be used for this distinction. First, a rough division of the models into the two overarching classes of exact match models (Subsection 3.3.1) and best match models (Subsection 3.3.2) is made. Within these classes, a further differentiation is then applied. First, the IR models that can be assigned to the class of exact match models are introduced. Then, the IR models that can be assigned to the class of best match models are introduced. A description of what characterizes each class, how the assignment of models to a class is organized, and which models can be assigned to each class will be provided in the following subsections.

3.3.1   Exact Match Models

The exact match models describe those IR models that are based on dividing the document collections to be searched exclusively into the two classes *relevant* and *not relevant*. All documents that *exactly* match all components of the query are considered relevant. Any document that does not fully or only partially meet the criteria formulated in the query is considered not relevant.

Examples of exact match models include many retrieval models that expect a query in the form of a formalized request [9]. This corresponds, for example, to those IR models that expect a query in the form of Boolean algebra. Boolean algebra allows certain combinations to be defined, which can be evaluated as either true or false, or in the case of IR, as relevant or not relevant. In IR, the terms of a query are combined using logical operators (OR,   AND, and NOT) through this formalism. The expression thus formed is then compared to the terms of a document. In this comparison, it is checked whether the query expression can be resolved to true with respect to the terms of a document, making the document

Table 3.1 Summary of the experiments

Document/Term Assignment	t_1	t_2	t_3	t_4	t_5
d_1	X		X		X
d_2		X	X		X
d_3	X		X	X	X
d_4	X	X	X		

relevant to the query. A detailed introduction to Boolean algebra is not provided here. Introductory materials on this topic can be found in many freely available publications [9].

The basic implementation of retrieval using Boolean expressions is briefly outlined schematically below. An artificial and simple small example is used to illustrate the concept in practice. It is assumed that a document collection D is given, with $D = \{d1, d2, d3, d4\}$. The individual documents are represented in Table 3.1 by the terms they contain t_1 to t_4 (an x indicates that a term is present in a document).

A Boolean expression that is a valid Boolean query q on this document collection is, for example, $q_1 = t_1$ AND t_3 or, expressed mathematically, $q_1 = t_1 \wedge t_3$. The OR function has the symbol $\vee$, the *NOT* function is represented by the symbol $\neg$.

An IRS that implements the exact match model then returns the documents d_1, d_3 and d_4 to the user. A ranking of the returned documents cannot be performed, since the selection of documents is based solely on checking the validity of the Boolean expression. Since only the exactly matching documents are returned, this class of retrieval models can also be referred to as **fact-based retrieval** [9]. Boolean retrieval is therefore often used in applications where users have a clearly defined idea of the desired search result. This is, for example, relevant in patent retrieval, i.e., in IR processes that search for documents containing a precisely specifiable set of information to demonstrate the uniqueness of a subject matter to be patented.

> Exact match methods always return exactly matching results.

3.3.2 Best Match Models

The exact match models described in the previous subsection divide the documents of a collection into the two classes relevant and not relevant with respect to the query. In addition to this rather restrictive selection property, another fundamental area of modeling in IR is the introduction of models that implement the classical, conceptual IRS model (see Subsection 3.2). The fundamental difference from exact match models is that these models provide for the output of retrieval status values (RSV). As already explained in Subsection 3.2, an RSV represents the relevance between a query and a document. This class of IR models is therefore also referred to as best match models. All best match models share the characteristic that they compute which documents best match a query.

Based on the RSV, a ranking is then performed when presenting the results. The ranking ranges from relevant to less relevant. When calculating the RSV, it is taken into account that users often have only a vague idea of the content of the data collection and cannot express their information need exactly. In other words, best match models attempt to deal as effectively as possible with an uncertainty factor or with the vaguely formulated information need (see also Belkin's ASK model from Subsection 1.4).

Over time, a large number of these best match models have been developed. The most important of these models are described in more detail in the following subsections. Here too, the models are grouped into chapters according to defined criteria.

3.3.2.1 Fuzzy Retrieval

Fuzzy retrieval is an extension of classical Boolean retrieval. The advancement lies in the ability not only to make the binary decision as to whether a document is relevant or not, but to introduce a gradation of the relevance of a document to the query. For this requirement, so-called fuzzy sets are used. The term fuzzy set refers to a concept introduced by Zadeh [10]. To begin, consider the definition of a simple set function on the set S. This function maps a value x based on the membership of $x \in S$ to a subset A. The subset A is thus defined by the function: $f_A : S \rightarrow [0, 1]$, where

$$f_A(x) = 1 \Leftrightarrow x \in A$$
$$f_A(x) = 0 \Leftrightarrow x \notin A$$

A fuzzy set now extends this binary mapping rule with mapping values from an interval: $i_B : S \rightarrow [0, 1]$, where B is a so-called label that describes an interpretation of the contained values. To illustrate this approach, let S be, for example, the set of all non-negative integers. It is further assumed that each element of this set S can be interpreted as a human age. Based on these assumptions, we can define the following function to identify people of young age:

$$I_B(x) = 1 \Leftrightarrow 0 \leq x \leq 20$$
$$I_B(x) = (50 - x) \leftrightarrow 20 \leq x \leq 50$$
$$I_B(x) = 0 \Leftrightarrow 50 \leq x$$

This function thus specifies that all people up to the age of 19 are relevant for this query. People aged between 20 and 50, on the other hand, have a decreasing relevance (in the interval [0,1]) with respect to the query. All people over 50 are defined as not relevant. This example is, of course, very specifically tailored to a query. More generally, the functionality can be described as follows: If a document d is contained in the set A to a degree of g_A and in a set B to a degree of g_B, then the following functions can be defined for the union, intersection, and complement:

Table 3.2 Example document and term distribution

Document/Term	t_1	t_2
d_1	0.5	0.8
d_2	0.9	0.1

$$g_{A \cup B} = \max(g_A, g_B) \tag{3.1}$$

$$g_{A \cap B} = \min(g_A, g_B) \tag{3.2}$$

We now take the simplified documents in Table 3.2 and assume the degree of membership of the terms t_1 and t_2.

From this distribution, it follows that for the term $t_1 = \{(d1, 0.5), (d2, 0.9)\}$ and for the term $t_2 = \{(d1, 0.8), (d2, 0.1)\}$ the following holds. The query $q = t1$ AND $t2$ is evaluated using the intersection and results in: $\{(d1, 0.5), (d2, 0.1)\}$. According to this method, document d_1 is to be considered more relevant than document d_2, and the documents can be sorted according to this criterion.

IR models that are based on the calculation of a distance assume that the relevance of a document to a query can be calculated based on the spatial proximity of a query representation and a document representation. The vector space model is a classic representative of this category and is therefore introduced first. Building on this, an extension of the exact match models based on the idea of the vector space model is also introduced.

It should also be mentioned here that, although distance metrics typically determine the distance between a query and a document, under the "query-by-example paradigm," a document itself can also serve as a query. In this case, the metrics would then calculate the distance between two documents, making it possible to sort the entire collection by similarity to a chosen element. Furthermore, the "query-by-example paradigm" also shows that both queries and documents are converted into an IR representation using the same type of indexing.

3.3.2.2 Vector Space Model

A fundamentally different retrieval model from the fuzzy model was developed in the 1960s as part of the development of the SMART IR system (SMART—System for the Mechanical Analysis and Retrieval of Text) under the direction of Gerard Salton at Cornell University [11]. The development of SMART is based on the idea that both document and query representations can be depicted as vectors in a space. In this so-called vector space model (VSM), the relevance of a query vector $\vec{q} \in Q$ to a document vector $\vec{d} \in D$ is determined by a measure of distance. The assumption is that the closer a document vector and a query vector are to each other, the more relevant they are to each other.

To represent a document or query as a vector, the VSM first determines the space of all possible terms from the entirety of all documents in a collection. Each of these n terms is then assigned a dimension in the vector space. A single

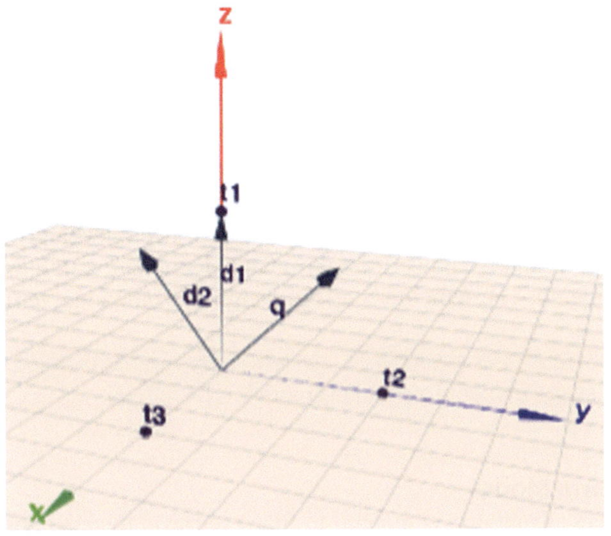

Fig. 3.6 Example vector space model

document or query can thus be defined by an n-dimensional vector. In the simplest case, a vector contains the value 1 at dimension x if the term represented by that dimension is present, or otherwise a 0. It is assumed that the vector space is orthogonal. This assumption means that all vectors representing terms are orthogonal and normalized [2, 11].

To better illustrate the approach, let the documents d_1 (contains only the term t_1), the document d_2 (contains the terms t_1 and t_3), and the query q (contains the terms t_1, t_2 and t_3) be given. In vector form, this means: $d_1 = (0, 0, 1), d_2 = (1, 0, 1)$, and $q = (1, 1, 1)$. The situation is illustrated in Fig. 3.6.

To calculate the relevance of a document to the representation of a query, the distance between the two vectors is then calculated in an appropriate way. The scalar product is often used for this calculation. When applied, if both vectors are perpendicular to each other, a value of 0 is calculated. If both vectors have the same direction, a maximum value is obtained (1 in a normalized space). Vectors that are perpendicular to each other are thus considered not relevant in the VSM, and vectors with the same direction are considered highly relevant. The similarity of two vectors, calculated using the scalar product and interpreted as the relevance of a query to a document, is calculated as follows:

$$sim(d_j, q) = d_q \cdot q = \sum_{i=1}^{t} d_{i,j} \cdot q_i \tag{3.3}$$

For our example, this yields $sim(d_1, q) = 1$ and $sim(d_2, q) = 2$. Document d_2 is thus more relevant than document d_1. Alternatively to the scalar product, relevance can also be calculated using the cosine. The smaller the angle between two vectors, the more relevant they are to each other. The cosine is calculated as:

$$\cos sim(d_j, q) = \frac{\vec{d_j} \cdot \vec{q}}{|\vec{d_j}| \cdot |\vec{q}|} \tag{3.4}$$

In addition to the simplified assumption that every term in the document database is equally important, numerous other methods have been developed that take into account the discriminative power of a term for a document within the entire document set considered. These methods assign weights to individual terms.

> The vector space model is a standard model in IR.

3.3.2.3 P-Norm Model

The VSM offers advantages compared to fuzzy retrieval models. For example, the geometric interpretation is easy to understand. However, a disadvantage is that the VSM no longer allows for structured queries. Queries that, for example, should include an OR combination to also consider synonyms of a term in the search cannot be represented. This disadvantage is addressed in the so-called P-norm model [2, 11]. It represents a compromise between the strict rules of a Boolean query and the disadvantages of unstructured queries in the VSM. Like the VSM, the P-norm model is based on introducing and calculating distance measures between the representation of a query and a document. The underlying assumption is that if a measure of distance between the two representations can be calculated, this distance can also be interpreted as a relevance value.

P-norms are a class of vector norms in mathematics, defined for real numbers $p \geq 1$. For the case $p = 1$ this is called the sum norm, $p = 2$ is the Euclidean norm, $p = \infty$ is the so-called maximum norm. All P-norms are equivalent, monotonically decreasing as p increases, and form the basic building block for norms of other mathematical objects such as sequences, functions, or matrices.

To introduce the P-norm model, let us first consider the simple case of a query with only two terms (t_1 AND t_2). The expectation would now be that all documents containing both terms are returned. However, one would also expect that in the case of an OR combination, the situation where neither term is present should be avoided. If you imagine this relationship graphically, with both terms mapped to their own axis of a Cartesian coordinate system, you get Fig. 3.7.

Overall, this leads to the requirement that for AND queries, found documents should be sorted in descending order of proximity to the point (1, 1), and for OR queries, in ascending order of distance to the point (0, 0). The distance to the point (0, 0) can be measured using the Euclidean distance as follows:

$$\sqrt{(d_{t1} - 0)^2 + (d_{t2} - 0)^2} \tag{3.5}$$

Fig. 3.7 Example P-norm model

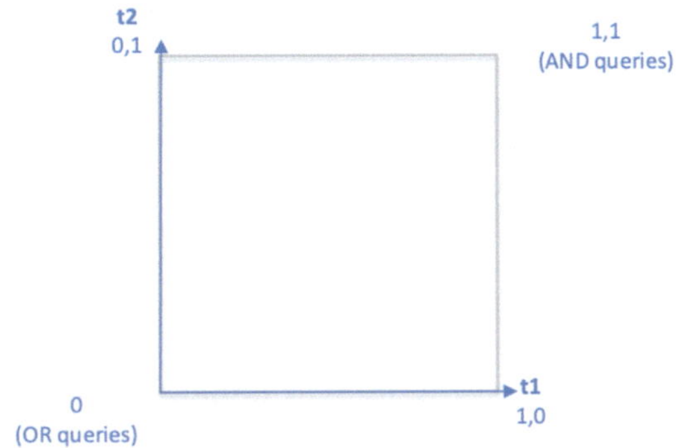

where d_{t1} denotes the weighted term t_1 in document d. The distance to the point $(1, 1)$ can analogously be determined by calculating the following formula:

$$1 - \sqrt{(1 - d_{t1})^2 + (1 - d_{t2})^2} \tag{3.6}$$

Further refinements and optimizations of the P-norm model are listed in [12] and are not discussed in this subsection.

3.3.2.4 Probabilistic Models

The models introduced so far are based on the evaluation of Boolean expressions or on the calculation of a distance measure, which allows conclusions to be drawn about the relevance between a document and a query [2]. The distance here refers to a measure of similarity between the representation of a query and that of a document. In addition, another branch of IR deals with probabilistic approaches to explaining and calculating the relevance of a document with respect to a query.

Before the models are presented in detail, a few basics of probability theory are introduced, which are fundamental for understanding the following models. In addition, the so-called probabilistic ranking principle (PRP) is introduced, which forms the basis of all probabilistic IR models.

3.3.2.5 Probability Calculation in IR

All probabilistic methods are based on the use of insights from probability theory, or probabilistics, a subfield of mathematics. A fundamental concept of probabilistics, which is also applied in IR, is the definition of independent and dependent events. To refresh and illustrate the concepts of independent and dependent events, we first refer to the classic examples of rolling a die and drawing balls from an urn.

First, it is clear that if a die is rolled x times in succession, the result of each roll is always independent of the number rolled previously. In contrast, if balls are drawn one after another from an urn and not replaced, then each subsequent draw affects the next, since the set of available balls changes. Such an event is therefore referred to as a dependent event.

The example of the balls is interpreted more abstractly in IR. Here, we speak of an event space consisting of the set $Q \times D$ of all possible queries Q and documents D in a collection. $P(A)$ then denotes the probability that an event from the space occurs, and the following holds: $0 \leq P(A) \leq 1$. To express that two events A, B occur together, the conditional probability $P(A, B)$ is defined. The case where both events are dependent on each other is expressed as: $P(A|B)$. This means the probability that event A occurs given event B. The multiplication rule shows the following relationship $P(A \cap B) = P(A|B)P(B) = P(B|A)P(A)$.

3.3.2.6 The Probabilistic Ranking Principle

The so-called Probabilistic Ranking Principle (PRP) is a theoretical approach to justify optimal retrieval in probabilistic IR models using a cost factor C. C denotes the cost associated with finding a document d_m for the query q_k. C^-, on the other hand, denotes the cost of finding any non-relevant document d_i in the data collection. The event space under consideration is as before $Q \times D$ and an element of the set is (d_m, q_k).

The fundamental assumption in PRP is that, for a given query, relevant documents should be found. According to Fuhr [5], this means for IR models based on probability calculations that, in perfect retrieval, all relevant documents are found before the non-relevant documents. In practice, this goal cannot be achieved, since the calculation of the RSV only works with representations of documents and queries. Therefore, realistically, only so-called optimal retrieval can be achieved. Optimal retrieval is achieved when all representations of documents have been sorted by the retrieval system in descending order of relevance. To theoretically underpin this statement, a decision-theoretic justification is chosen.

First, it holds that an IR model should prefer relevant documents over non-relevant ones. Thus, it is established that the cost of retrieving a non-relevant document is higher than the cost of retrieving a relevant document: $C^- > C$. The expected cost C_E of retrieving a document d can also be represented as:

$$C_E(d) = C \cdot P(R|q, d) + C^- \cdot (1 - P(R|q, d)) \tag{3.7}$$

To keep retrieval costs as low as possible, relevant documents must therefore be ranked before non-relevant documents. Otherwise, increasing costs are expected and the retrieval is not optimal.

3.3.2.7 Probabilistic IR Models

IR models that can be interpreted probabilistically can, among other things, be distinguished according to two approaches: the model-oriented approach and the description-oriented approach [1]. The model-oriented approach is based on independence assumptions regarding the elements of a document or query representation. The description-oriented models, on the other hand, are based on a document or query representation that relies on the description of certain features (e.g., the frequency of co-occurrence of terms: term co-occurrence). Furthermore, classical retrieval models can also be represented using propositional logic, which allows for a probabilistic interpretation. As a representative of all possible probabilistically interpretable IR models, a model-oriented approach will be presented in this chapter.

Probabilistic IR models that belong to the class of relevance models are based on certain independence assumptions. The Binary Independence Model (BIM) is a classic representative of this category and will be introduced as an example. The presentation follows [13]. First, it should be reiterated that the goal of BIM is to calculate the probability that a document d is relevant. As mentioned at the beginning of this subsection, this calculation can be based on the formulation of conditional probability. If the conditional probability is applied to the statement just made, then the following holds:

$$P(R|d) > P(R^-|d) \tag{3.8}$$

The problem with this calculation is that resolving $P(R|d)$ cannot be performed without given probability distributions. Therefore, let us first assume that for a query there is a set of relevant and a set of non-relevant documents, so that at least $P(d|R)$ can be calculated. Suppose, in this case, the information is available that in the set of relevant documents, the probability for the word "president" was calculated as 0.02 and for the word "Lincoln" as 0.03. In this case, it would initially hold that $P(A \cap B) = P(A|B)P(B)$. However, taking into account the independence assumption of terms, the following relationship holds: $P(A \cap B) = P(A) \cdot P(B)$. For a new document in the set of relevant documents, the probability $P(d|R) = 0.02 \cdot 0.03 = 0.0006$ would thus result.

At this point, it is still unclear how the probability $P(R|d)$ can be calculated. For this, the following Bayes' rule is introduced:

$$P(R|d) = \frac{P(R|d) \cdot P(R)}{P(d)} \tag{3.9}$$

$P(R)$ denotes the a priori probability that a document is relevant, and $P(d)$ acts as a normalization factor. This allows us to formulate the following classification rule for documents, regarding the division into the classes relevant and non-relevant: $P(d|R)P(R) > P(d|R^-)P(R^-)$.

The left side of the equation is also referred to as the likelihood ratio. However, the goal of BIM is not classification, but the calculation of an RSV. For this, we continue to assume that documents are represented by the set of terms they contain. The representation of a document can then be made using a binary vector $\vec{x} = (x_1, x_2, \ldots, x_n)$ that includes all terms occurring in the collection. This means: if a term is present in a document, it is marked as 1 in its vector representation. If the term is not present, it is marked as 0 in the vector.

In addition to the introduced form of representation, BIM also concerns the so-called cluster hypothesis. The hypothesis states that the set of all terms $T = t_1, t_2, \ldots, t_n$ in the entire collection under consideration is distributed differently in relevant and non-relevant documents. Finally, the independence assumption is made that, even considering the cluster hypothesis, the terms in the set T are independent of each other.

However, to calculate the likelihood ratio, solutions are still needed to compute $P(d|R)$ and $P(d|R^-)$. BIM now introduces a simple procedure for this calculation. Due to the independence assumption of the available terms, an estimate for $P(d|R)$ can also be calculated as the product of the individual term probabilities: $\prod_{i=1}^{t} P(d_i|R)$ or $\prod_{i=1}^{t} P(d_i|R^-)$.

So far, sets of relevant and non-relevant documents have been considered. However, what has not yet been taken into account is that these sets only arise with respect to an information need, which is expressed through a set of words. The query is the only source of information by which we can determine the set of relevant terms in a document. All other terms that may occur in the collection are initially considered equally relevant. Thus, they have the same occurrence probability ($p_i = s_i$). In this case, the summation is only performed over the terms that are present in both the query and the documents. For a given query, a value results that simply corresponds to the number of terms that appear in both the document and the query. If no further information is available, p_i can be assumed to be constant and s_i can be estimated. For p_i, for example, a value of 0.5 could be assumed, and for s_i, an estimate could be the ratio between the number of documents containing the ith term and the total number of documents in the collection. If information is available about the frequency of occurrence of terms in relevant and non-relevant documents, these can be summarized in a so-called contingency table. One method to obtain such information is, for example, the so-called relevance feedback method. This evaluates user feedback on the relevance of documents from an initial search run. A detailed description of the method is given in Subsection 3.3.3. If relevance feedback data is available, it can be integrated into the calculation models mentioned above.

> Best-match methods provide ranked result sets.

3.3.3   Relevance Feedback

Up to now, the calculation of document relevance for queries has generally been considered a self-contained process. An exception to this is the BIM, which requires information for calculating an RSV that is not available during an initial search. Relevance feedback refers to a group of methods that introduce a certain iterative character into the calculation of relevance. This is achieved by taking into account relevance calculations from previous searches in a current search. In this sense, relevance feedback is also a retrieval optimization technique, based on the assumption that the outcome of a current IR process can be improved by relevance judgments on the search results of a previous search.

There are various approaches to implementing relevance feedback. What they all have in common is that they extract information from one or more previous IR processes. This extracted information is then used to optimize the result of the current search. In general, this optimization is based on either a modification of the query or a reordering of the search results. The modification of the query is then based either on an expansion of the elements of a query or a reweighting of the terms in a query [1, 14, 15]. Some possible forms of relevance feedback will be presented in the following subsections.

3.3.3.1 Explicit Relevance Feedback

In explicit relevance feedback, the optimization of a search result is based on the explicit relevance judgment of users of an IRS regarding a certain number of documents marked as relevant. When applying explicit relevance feedback, the IRS must provide users with the opportunity to review the results of a search for relevant hits and to mark these with a relevance judgment. The assessment of the relevance of these documents can take various forms. Possible relevance judgments are relevant, irrelevant, neutral, or the assignment of a real number within a certain interval (e.g., from the interval [0,1]). The advantage of explicit relevance feedback is that only those documents that users have actually found relevant to their query are selected for further calculation in the relevance feedback process. Explicit relevance feedback is therefore less susceptible to noise or misjudgments than implicit and pseudo-relevance feedback, which will be presented in the following subsections. However, users must accept the effort of examining a certain number of documents returned by the IRS for their actual relevance.

In general, explicit relevance feedback proceeds according to the following scheme:

1. Submission of the query, e.g., in the form of a vector $\vec{q_m}$
2. The IRS returns a set of relevant documents.
3. Users provide a relevance assessment of the returned documents in the form of feedback data.
4. The IRS optimizes the query vector $\vec{q_m}$ based on the feedback data and thus generates $\vec{q}\,'$
5. The retrieval process is carried out with the optimized vector $\vec{q}\,'$.
6. If necessary, repeat steps 2–4.

To illustrate the approach, the Rocchio method [16] will be described. The Rocchio method uses explicit relevance feedback in the VRM. In the VRM, the similarity between a document vector $\vec{d_m}$ and a query vector $\vec{q_m}$ is determined using a similarity measure, such as the scalar product. The Rocchio method is based on the idea of considering relevant and non-relevant documents as two distinct sets. In this case, it must be possible to modify the query vector through a calculation so that it tends to move further away from the non-relevant set and toward the center of the relevant document set. The main principle of the Rocchio algorithm is to change the weights of the terms in the search query based on user feedback. Terms that occur frequently in relevant documents receive a higher weight, while terms that occur frequently in non-relevant documents receive a lower weight. In this way, relevant terms are reinforced, while non-relevant terms are weakened.

> With explicit relevance feedback, the user provides relevance judgments.

3.3.3.2 Implicit Relevance Feedback

In the application of implicit relevance feedback, the aim is to infer the relevance of documents without using explicit relevance judgments from users. For this purpose, among others, [17] draw on users' interaction data with the IRS and use these implicitly recognized insights into search behavior as input for a relevance feedback method. Such information is observed, collected, and used to optimize the search result without the users' awareness during the search process. The weakness of implicitly collected relevance feedback data lies in the fact that the actual relevance of the collected data can only be inferred with a certain probability. The inaccuracy results from the uncertainty as to whether the collected interaction data were actually relevant to the search process or not. Since no explicit judgment is requested from users, a residual uncertainty remains. The difficulty in applying implicit relevance feedback therefore lies particularly in collecting only truly relevant information. The major advantage of implicit over explicit relevance feedback, however, is that there is no additional effort required from users during retrieval: users do not have to read and assess documents.

> Implicit relevance feedback uses user interactions indirectly for relevance assessment.

Some methods that have been used to implement implicit relevance feedback are presented in the following subsections. These are based on findings from research conducted in the context of Internet search engines. The following chapters are particularly based on findings by Joachims [18].

3.3.3.3 Click-Rate Data

The so-called click rate of users records IRS interactions via mouse clicks. The click-rate data thus collected can be described in the form of a triple $Clickrate\text{-}Data = (q, ro, c)$, where:

- q: current query
- ro: ranking of the result of a search session
- c: set of links clicked by users in the result

Click-rate data are particularly interesting because users generally do not click on links at random. Users activate a link because they have formed some judgment about the relevance of the information that might be behind a link. Click-rate data can be recorded relatively easily and without active participation from users during a search session. They are therefore particularly well suited as implicit relevance feedback data. Joachims [18] especially emphasizes the importance of information that arises from the relationships between the individual components of the click-rate data triple. These relationships correspond to the dependency of the ranking on the query (resulting from the implementation of the search engine) and the dependencies of the set c, which depends on both the query and the ranking of the results.

However, a problematic aspect is the fact that the number of search results in most search engines is so large that users only look at a fraction (usually only the first 10 results) of the search results. The probability that users will look at the result in position 1000 is vanishingly small. Because of this, it is not possible to establish an absolute scale for the relationship between c and ro.

Another problem arises from the fact that users read the result list of search hits from top to bottom. The order of clicks therefore cannot be an indicator of a relevance ranking of the clicked links. A study co-authored by Joachims [18] also concludes that click-rate data contain a lot of useful information, but can be error-prone. In this study, eye trackers were used to draw conclusions about user behavior during search.

Not only the order of search results influences clicking behavior. Users also judge based on the quality of all textual snippets of a result. As a result, click-rate data must be evaluated relative to their arrangement and relative to the other search results. Furthermore, click-rate data can be interpreted in different ways. Therefore, possible strategies for interpreting the data will be presented here. These are also based on the publication by Joachims [18]. Figure 3.8 further illustrates these.

Strategy 1 (Click > Skip Above): Represents a preference ordering of the search results by users. For a given ranking $(l_1, l_2, l_3, \ldots)$ and a given set C containing the links clicked by users, a preference order can be extracted as follows: $rel(l_i) > rel(l_j)$ for all pairs $1 \leq j < i$ with $i \in C$ and $j \notin C$.

Strategy 2 (Last Click > Skip Above): Considers only the last click to create a preference: if $i \in C$ is the last clicked link, then a preference order can be extracted as follows: $rel(l_i) > rel(l_j)$ for all pairs $1 \leq j < i$ with $j \notin C$.

Strategy 3 (Click > Earlier Click): Is based on the assumption that users are more informed during later clicks than earlier ones. Therefore, it can be assumed that later clicks are also more relevant clicks. Let $t(i)$ with $i \in C$ be the time at which link l_i was clicked. A preference order can then be extracted as follows: $rel(l_i) > rel(l_j)$ for all pairs j and i with $i, j \in C$ and $t(i) > t(j)$.

Strategy 4 (Last Click > Skip Previous): The text fragments of search results that are most reliably viewed by users are those directly above a clicked link.

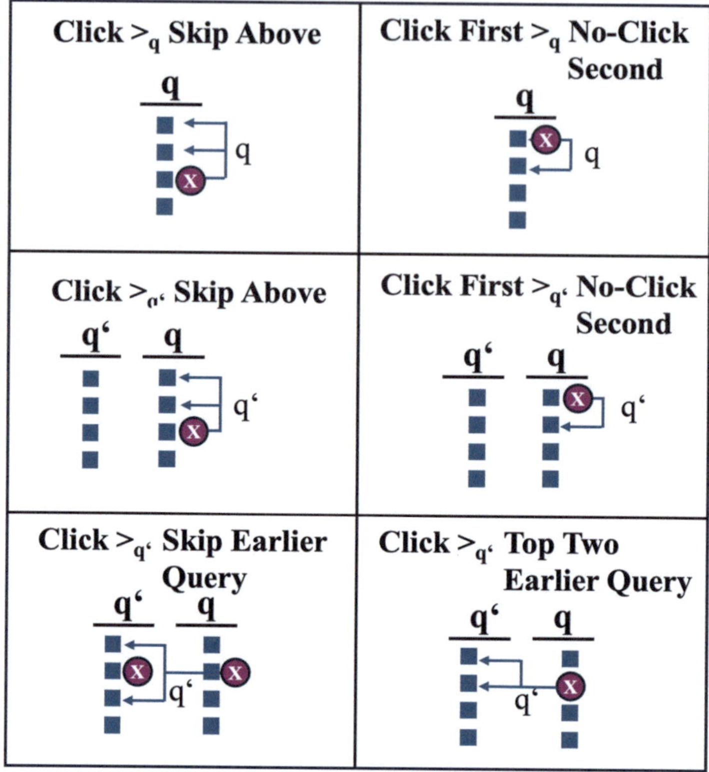

Fig. 3.8 Relevance feedback strategies according to [18]

A preference order can in this case be extracted as follows: $rel(l_i) > rel(l_{i-1})$ for all pairs $i \geq 2$ with $i \in C$ and $i - 1 \notin C$.

Strategy 5 (Click > No-Click Next): An evaluation showed that the text fragments immediately following a clicked link are usually also read. Therefore, a preference order can be extracted as follows: $rel(l_i) > rel(l_{i+1})$ for all $i \in C$ and $(i + 1) \notin C$. This represents a condition between a clicked link and an immediately following non-clicked link.

Furthermore, additional strategies for interpreting click-rate data are conceivable [19]. Unlike those of Joachims, these strategies are based on query chains. Query chains refer to the grouping of multiple queries with the same information need. This corresponds to repeatedly reformulating a query. The advantage of this approach is that the relevance assessment of search results is based on a much larger number of result documents. This is in contrast to [18], where only results from a single query are considered. This approach therefore includes significantly fewer documents to draw upon.

Strategy 6 (Click First > q No-Click Second): Is based on the fact that users typically look at the first results of a search result (before clicking). From this, it is inferred that if the first search result is clicked but not the second, the first is more relevant than the second.

Strategies 7 and 8 (Click > q_0 Skip Above and Click First > q_0 No-Click Second): Is based on the assumption that both queries q and q' belong to a search with the same information need. If users had received the results of q' for query q, they would have preferred to have the search results from q' above those that were skipped.

Strategy 9 (Click > q_0 Skip Earlier Query): States that a search result in q is more strongly preferred by users than any search result from an earlier query q'.

Strategy 10 (Click > q_0 Top Two Earlier Query): In the case where no document was clicked in an earlier query q', the fact is exploited that users usually look at the first two results of a search. Thus, the first two results are preferred.

The list of possible strategies presented here also highlights a typical problem in practice: for the effective use of click-rate data, it is necessary to find a strategy for interpreting a click that is appropriate for the application. There is therefore no universally valid standard procedure. Instead, applications must be individually optimized for typical use cases and, if necessary, for typical user groups.

3.3.3.4 Reading Time, Scrolling, Interaction

Research interest in extracting implicit relevance feedback data also extends to reading time, scrolling, and further user interaction with the IRS [20]. The evaluation of these relationships is based on the following three hypotheses:

- Users spend more reading time on documents they find relevant than on documents they do not find relevant.
- Users scroll more in documents they find relevant.
- Users interact more with documents they find relevant.

The result of the study is that reading time, scrolling, and interaction only have real significance in certain environments (laboratory conditions). In this research, reading time, scrolling, and interaction did not achieve the desired significance. This conclusion is drawn from the fact that the reading time for relevant and non-relevant documents was comparably high. The same result was found for scrolling effort and interaction with the IRS.

3.3.3.5 Pseudorelevance Feedback

Pseudorelevance feedback is based on the assumption that the first n documents returned by an IRS for an initial query are also judged relevant by users. Based on this assumption, the first n documents sorted by the IRS's ranking method are used to optimize the result of a search. Pseudorelevance feedback can also be seen as a special case of implicit relevance feedback. In this case, the IRS's ranking method is primarily used to select the implicit feedback data. Pseudorelevance feedback is relatively easy to implement, since the source for relevance judgments is predetermined and already integrated into the IRS.

3.3.4   Query Optimization through Modification of the Query

Formulating an information need is not a simple task due to its intrinsic vagueness. Query optimization methods provide support for refining the query. In general, query optimization involves reweighting or expanding query terms, which are, for example, derived from a set of initially relevant documents. However, this raises the question of which terms from identified documents are actually suitable for expanding the query and in what form. The following subsection discusses methods that describe corresponding solutions.

For further literature, see references [21, 22]. Furthermore, term co-occurrence is introduced as a means to determine and evaluate expansion terms. First, it should be generally noted that the same insights apply to the initial selection of expansion terms as are also evident in the weighting of index terms:

- Very high-frequency terms are not helpful for expansion.
- Terms with medium frequency have a positive impact.
- Low-frequency terms can contribute to improved results in exceptional cases.
- Very low-frequency terms can have a very positive impact on the search result.

> Query optimization refines the formulation of the information need.

3.3.4.1 The F4 Algorithm

The F4 algorithm is a probabilistic ranking method [21]. It is based on the following independence assumption: the distribution of terms in relevant documents is independent, and the distribution of terms in non-relevant documents is also independent (clustering hypothesis). This is to be understood in contrast to the assumption that the distribution of terms in relevant documents is independent, as is the term distribution in all documents.

Term independence here means that the occurrence of a particular term is independent of the occurrence of another term. The crucial difference between the two assumptions lies in the set of documents to be considered. This is due, on the one hand, to the collection as a whole and, on the other hand, to the probability difference of a term's occurrence with respect to the non-relevant documents. Furthermore, the F4 algorithm assumes an ordering principle:

The relevance probability of a term is based both on the presence of search terms in documents and on their absence in documents.

In other words, the method is based on the probability with which a term appears in a relevant document, in combination with the probability with which it appears in a non-relevant document. The formula for calculating the RSV of a term t using F4 is as follows:

$$F4 = \log \frac{r \cdot (N - n - |D^R| + r)}{n - r + r(|D^R|) - r} \tag{3.10}$$

To avoid undefined states in practice, a value of 0.5 is added to each term in the formula:

$$F4 = \log \frac{(r + 0.5) \cdot (N - n - |D^R| + r + 0.5)}{n - r + r + 0.5(|D^R|) - r + 0.5} \tag{3.11}$$

Here,

- N is the number of documents in the collection,
- $|D^R|$ is the number of relevant documents,
- n is the number of documents containing the term t,
- r is the number of relevant documents containing the term t.

Analyses have shown the tendency of the F4 algorithm to assign a high ranking to terms with low frequency in the collection—just as terms with low frequency in the set of relevant documents are given a high ranking.

3.3.4.2 WPQ

WPQ calculation was introduced by Robertsen in [23] and has been further developed several times. It consists of a weighting function that takes as input the difference between the parameters p_t (probability that t appears in a relevant document) and q_t (probability that t appears in a non-relevant document), resulting in the following equation:

$$wpq = q_t(p_t - q_t) \tag{3.12}$$

For weighting the difference, Efthimidias and also White [24] propose the use of the F4 method:

$$wpq = \log \frac{(r)/(R - r)}{(n - r)/(N - n - R + r)} \cdot \left(\frac{r}{R} - \frac{n - r}{N - R}\right) \tag{3.13}$$

Here,

- r is the number of seen relevant documents containing the term t,
- n is the number of documents containing the term t,
- R is the number of seen documents for q,
- N is the number of documents in the database.

3.3.4.3 Term Co-occurrence

Term co-occurrence describes a measure of the association between two terms. The assumption is that two terms that appear together in a text (sentence, paragraph, abstract, document, etc.) have a certain connection (association) to each other. If such an association is not limited to a single text in the document collection but is also found in other texts within the collection, it can be assumed that there is a stronger association between the terms. Terms that do not appear together in any text of a collection have no co-occurrence. Van Rijsbergen provides the following association hypothesis [25]: "If an index term is good at discriminating relevant from non-relevant documents then any closely associated index term is also likely to be good at this."

The pairwise association between terms can be represented in a term co-occurrence matrix. In such a matrix A, the entry $A_{x,y}$ reflects the strength of association between term x and term y. This process is not limited to the association of two terms but can be extended to associations among any number of terms.

However, when using term co-occurrence, it should be noted that the co-occurrence of terms that appear very frequently in the text collection is particularly high. Highly frequent terms, in turn, discriminate between different texts only very weakly. Under these circumstances, they cannot distinguish well between relevant and non-relevant documents. Considering term co-occurrence in the context of multimedia objects, it is evident that many additional, automatic connections can often be generated. In image recognition, for example, detected objects can be related not only by the fact that they are present in the same image (or document), but also by their spatial positioning (above, below, left, right, in front of, behind). Such relationships can be represented by additional terms and thus also utilized through additional term co-occurrence measures.

> There are numerous methods for optimizing queries and results.

3.3.5 Software Libraries

This section introduces three common and representative software libraries that are frequently used in the context of indexing and retrieval. However, these serve only as exemplary representatives. It should be explicitly noted that, of course, there are many other libraries available.

3.3.5.1 Apache Lucene

Apache Lucene is a platform-independent IR library for full-text search and is freely available for download. Apache Lucene is an open-source project of the Apache Software Foundation (ASF), originally developed by Cutting in the mid-1990s. In 2001, the library became part of the Jakarta Project and in 2005 was promoted to a top-level project of the ASF [26]. Lucene was originally implemented in Java, but integration options are now available for other programming languages such as C, C++, C#, Ruby, Perl, Python, and PHP [26]. Apache Lucene can be used as an IR library wherever textual content is available that needs to be made searchable. The format of the text documents and the language used in the documents are not relevant, so any type of textual data can be processed with Apache Lucene.

To make documents searchable, Apache Lucene essentially performs two steps: First, prior to actual queries, an index is created from the given data set, with each document in the data set being indexed. This Lucene index is comparable to an index at the end of a book, which assigns page numbers to important terms. For given queries, Lucene then accesses this index and returns documents matching the query, weighted by relevance, as results. Classic information retrieval techniques are used as search algorithms. To determine the relevance of documents with respect to a query, Apache Lucene by default uses the vector space model [27]. Thanks to the preliminary indexing step and the intelligent use of search algorithms, Apache Lucene is highly effective and can therefore be used very efficiently even on large data sets. For this reason, Apache Lucene has become widely used as an IR library in recent years and is employed by, among others, Netflix, LinkedIn, and Atlassian Jira. When using Apache Lucene, it should be noted that it is merely a library and not a fully-fledged search engine. If additional components are required for the specific application, such as a runtime server, a user interface, crawling functionalities, or administration options, these must be added. However, with Apache Solr and Elasticsearch, there are two search engines available that are built on Apache Lucene and extend the library with additional functionalities.

3.3.5.2 Apache Solr

Apache Solr is a Java-based open-source search platform built on the IR library Apache Lucene. The framework was developed in 2004 as part of an internal project at the media company CNET by Seeley and donated to the Apache Software Foundation in early 2006 [27]. Today, Apache Solr is available as open-source software on the internet.

The Apache Solr search server enables the implementation of complex and scalable search solutions and extends Apache Lucene with features such as highlighting, spellchecking, auto-suggest functionality, and result grouping via faceting. In contrast to Apache Lucene, Apache Solr is a stand-alone server that can be used entirely without programming knowledge and comes with an administration interface that can be accessed via a browser, allowing for convenient management [27].

For indexing documents in various file formats such as XML, JSON, or CSV, Apache Solr already provides predefined structures. In the case of XML files, the files to be indexed must have the following format: The content of an XML file is enclosed by an <add> tag, which instructs the system to add the file's content to the index. Within this <add> tag, any number of <doc> tags can be inserted, marking the beginning and end of a document. The content of these <doc> tags, in turn, consists of any number of <field> tags. Each of these fields has a name and a value. An example of such an XML file is shown in Listing 3.1:

```
1  <add>
2          <doc>
3                  <field name="id">98223</field>
4                  <field name="name">MacBook Pro</field>
5                  <field name="description">A notebook for power
                       users.</field>
6          </doc>
7          <doc>
8                  <field name="id">1783</field>
9                  <field name="name">ThinkPad 650e</field>
10                 <field name="description">The Lenovo notebook
                       of the ThinkPad
11                 edition with notebooks for office solutions.</
                       field>
12         </doc>
13         <doc>
14                 <field name="id">29933</field>
15                 <field name="name">iPhone</field>
16                 <field name="description">The famous Apple
                       smartphone.</field>
17         </doc>
18  </add>
```

Listing 3.1 Example of a Solr-Index

To index XML files formatted in this way, some configuration settings must be made in Apache Solr. The two most important configuration files are solrconfig. xml and schema.xml. The schema file contains the schema definition. It specifies which fields are allowed, what types these fields have, and optionally defines a weighting strategy for them [27]. The solrconfig file, on the other hand, contains the specific configurations for the respective application, such as request handling, response formatting, caching, and more.

Once the documents have been indexed, they can subsequently be searched, with a query being submitted either directly via the Solr administration interface or via an HTTP request. Solr's response to a query consists of a response header and a response body. The header contains meta-information about the request, such as status and processing time. The body lists the documents matching the query, sorted by relevance. An example response to the query "notebook" is shown in Listing 3.2:

```
1  <response>
2          <lst name="responseHeader">
3                  <int name="status">0</int>
4                  <int name="QTime">115</int>
5                  <lst name="params">
6                          <str name="q">notebook</str>
7                  </lst>
8          </lst>
9          <result name="response" numFound="2" start="0">
10                 <doc>
11                         <str name="id">1783</str>
12                         <str name="name">ThinkPad 650e</str>
13                 </doc>
14                 <doc>
15                         <str name="id">98223</str>
16                         <str name="name">MacBook Pro</str>
17                 </doc>
18         </result>
19  </response>
```

Listing 3.2 Example of a Solr-Response

3.3.5.3 Elasticsearch

Elasticsearch is a distributed and scalable open-source search server for real-time search and analysis of documents, which, like Apache Solr, is built on the Lucene library. Although the search server, implemented in Java, does not have a long history since its initial release in 2010, it is already more widely used today than Apache Solr. Currently, Elasticsearch is available for download in version 2.3.3. Elasticsearch and Apache Solr offer similar functionality but focus on different aspects [27]:

- Apache Solr focuses on text search and the relevant functions for this, while Elasticsearch also provides various options for analysis and monitoring.
- Since its initial release, Elasticsearch has relied on a distributed architecture. Solr only followed suit with version 4.0 as part of SolrCloud.
- Both search engines use REST endpoints, with Elasticsearch exclusively allowing JSON format for incoming and outgoing data. Solr, on the other hand, supports various formats such as XML, JSON, and CSV.
- Elasticsearch offers a feature that allows the application to be automatically notified when new documents matching specified conditions are found. Solr does not support these so-called push queries.

In summary, both Apache Solr and Elasticsearch are very powerful open-source search engines, both suitable for indexing and searching large document collections. Whether Apache Solr or Elasticsearch is better suited for a specific software application depends primarily on the concrete requirements of the software to be developed and must therefore be evaluated on a case-by-case basis.

Now that a number of models and concepts have been introduced, the following subsection addresses an important topic: the unique and persistent identification of multimedia objects. This aspect has often been implicitly assumed so far, but in multimedia information systems it must be implemented with the utmost diligence. Therefore, persistent identifiers and the technologies, concepts, and models associated with them are now introduced.

3.4 Persistent Identifiers

The term Persistent Identifier refers to an association between a name and a resource. The naming of objects plays a fundamental role in the automated processing of digitally encoded documents and media objects, particularly with regard to their addressing. While operating systems typically recognize file names, user and process identities, networks generally add naming schemes such as Ethernet addresses, IP addresses, DNS names, and URLs. In heterogeneous networks, the number of types of identifiers increases even further. In principle, when searching for a persistent identifier, one could simply draw from this pool. However, the problem is that the Internet environment is characterized by constant change. New web servers go online every day, others—sometimes unnoticed—go offline. New formats and applications emerge, and old ones fall into oblivion. To achieve a reasonably reliable and, above all, long-term secured access to digital data on the Internet, additional concepts and methods are therefore required, which will be described in the further course of this chapter. However, one important remark should be made in advance, on which experts agree: An identifier scheme can at best support persistence, but never enforce it, because ultimately persistence is as much a technical as it is a social problem.

3.4.1 Requirements

With regard to properties that would be advantageous for a mechanism to implement a persistent identifier, the following criteria can be named [28]:

- **Uniqueness:** An identifier should never be able to name two different objects. It thus establishes a (partial) function between names and objects.
- **Injectivity:** Two different identifiers should name different objects.
- **Format supported indefinitely:** Any identifier system is useless if the format for the identifier is eventually lost.
- **As technology-independent as possible:** Technologies are known to be subject to rapid change. It would therefore be naive to believe, for example, that an operating system will still have the same basic functionality in 100 years as it does today. If the implementation of an identifier method relies on such assumptions, the method will become obsolete along with the underlying technology.
- **Independent of storage location:** If the storage location is part of the identifier, it becomes invalid when the location changes (see, for example, DNS names in URL addresses).
- **Semantic neutrality is necessary:** If an identifier reflects any kind of semantics, such as ownership of the resource, it becomes at least misleading when these change (e.g., in the case of a sale).
- **Interoperability of the format:** Until a uniform, universally valid identifier format exists, the format should easily be able to include other formats or be included by others.
- **Language support:** Identifiers should be able to handle languages that are not easily represented in ASCII (e.g., Chinese).
- **Ease of creation:** The effort required to create an identifier name and assign it to a resource should be as small as possible.
- **Simplicity and efficiency in resolution:** The effort required to go from the name to the resource pointed to by the identifier should be as small as possible.
- **Fault tolerance:** If the server managing the mapping of the identifier to the resource fails, another should be able to take over; otherwise, the impression arises that the identifier name is invalid. This can be ensured, for example, through redundancy.

- **Security:** Changes should only be possible by authorized persons.
- **Metadata:** It should be possible to retrieve information about the referenced object.
- **Granularity:** It would be beneficial if the access resolution with respect to the resource could be graded.

All approaches to implementing identifiers have in common that they introduce a level of indirection, thereby decoupling the identifier from the storage location.

3.4.2   Relevant Approaches and Implementations

When considering already existing mechanisms for a persistent identifier on the Internet, the Universal Resource Locator (URL) comes to mind first, which every Internet user deals with daily. URLs are described in **url** and form a uniform scheme for naming objects on the Internet. In the context of our discussion, one could, in simplified terms, take the position that URLs bind physical storage locations to names.

A URL, more precisely, consists of the components shown in Fig. 3.9. These include the protocol (such as http or ftp), the DNS name (like www.google.de), the port (80 is the standard www port), and the physical path. The properties of these components result in two major disadvantages when using URLs as persistent identifiers. If the resource is moved to another location (even on the same server), the address changes, and thus the identifier becomes invalid. This problem could be solved by automatically using server redirects. The second and more significant disadvantage is related to the DNS names in the middle part of the URL. The problem here is that the administrative rights in the DNS system, for use as a persistent identifier, are simply in the wrong hands, so to speak, because the DNS system can be simplified as a tree structure with several levels of address space name domains, which are in the hands of various owners and administrators. At the top is the top-level domain. Below that are the second-level domains such as "com", "org", "de", etc. The resolution of a DNS name is performed from right to left and thus corresponds in the graph to a traversal from top to bottom. Thus, resolving URLs results in a distributed interaction of a system of servers, each managing its own domain. The identifier owner therefore has no or only partial access to the management and resolution of the individual name components of

Fig. 3.9 Components of a URL address

Fig. 3.10 An example PURL and its components

their identifier. Instead, the resolution and management rights lie with the operator of the respective DNS servers and thus not, as would be necessary for simple persistent maintenance of mappings from names to resources, in a single hand.

The simplest attempt to make a URL quasi-persistent can be seen in the PURL project (Persistent URL) of the OCLC. However, the system's inventors do not consider it a final solution to the persistence problem for identifiers. Rather, it was designed so that a fairly straightforward transition to Unified Resource Names (URN) would be possible. The main idea is to add an additional layer of indirection to a normal URL by making the first name component, i.e., the DNS part of the PURL address, consist of a generally known resolver address (see Fig. 3.10). The resolver is simply a server that resolves the second part of the symbolic address of a resource into the actual physical location of the sought digital object. This second part of the address, as with URLs, consists of domains and visually resembles a Unix path address. Internally, the domains are also mapped to tree-like structures, as with URLs. The most important point is that now the PURL owners (called "maintainers") have access rights to all name components of their identifier.

The system is implemented using the server redirect feature, which is built into every web server by default, so the technical effort required to realize the system is not very high. Figure 3.11 shows the process of resolving a PURL address.

The resolver (or PURL server) takes the second part of the address and converts it into a URL. In the second step, the client can then access the resource in the usual way. It is now common practice to use one's own domain name in reverse order as a kind of unique identifier. This can be found, for example, in the identifiers for Java packages or libraries, in repositories for APIs or source code such as Maven, or in the provision of web services via UDDI. Another example, specifically from the field of archiving, is the Archival Resource Key system (ARK) of the California Digital Library. Compared to the two implementations presented so far, ARK offers significant enhancements without requiring much technical effort. Like PURLs, ARK identifiers (also called ARKs) work with standard web browsers, and on the server side, regular web servers are converted into ARK servers using server rewrite rules. The ARK system also uses an additional level of indirection to improve the persistence of its identifiers. In addition, the ARK system introduces two operators, so that with an ARK identifier one can access resources, access metadata using an additional operator (by appending a "?" to the identifier), and access the Promise of Stewardship (by appending "??" to the identifier). The Promise of Stewardship can convey the (expected) duration for which the object

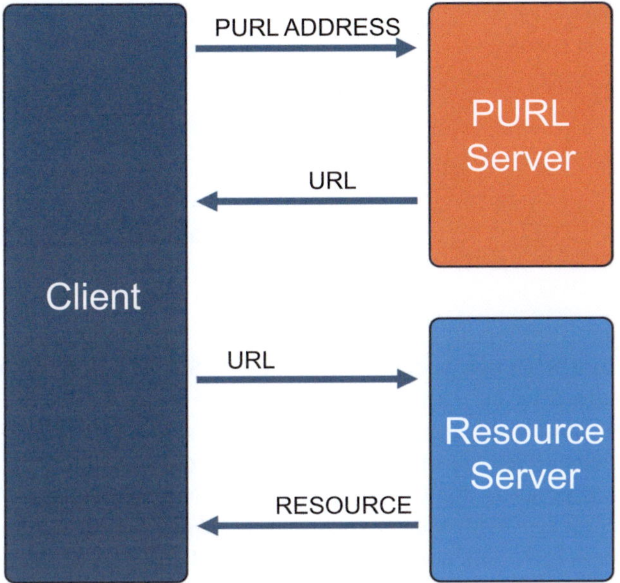

Fig. 3.11 Resolution of a PURL address

will remain accessible, how quickly the object is (expected to) change, and the (expected) duration for which the identifier itself will exist. ARK uses so-called Name Assigning Authorities as the highest administrative bodies in the ARK hierarchy, which are contacted when an identifier is to be assigned to a resource. In this way, hierarchies for the assignment and determination of unique ARK identifiers can be established and used.

Uniform Resource Names (URN) are an IETF standard for persistent identifiers, defined in RFCs 1737 and 2141. The following properties are required of an encoding syntax for a URN:

- Global scope (every identifier has the same meaning everywhere)
- Global uniqueness
- Persistence (URNs are intended to be valid forever)
- Scalability (it should not be possible to run out of URNs)
- Easy integration of other schemes into URN
- Good, efficient resolution to URLs

Point three of these requirements stands out in particular, as it contrasts with the basic assumption of the ARK implementation, namely that an identifier cannot be made persistent by technical means alone.

As shown in Fig. 3.12, the Namespace Identifier (NID), similar to the naming schemes discussed earlier, forms the top-level name component. A specialty of the URN concept is that a separate Name Space Syntax (NSS) can be specified for

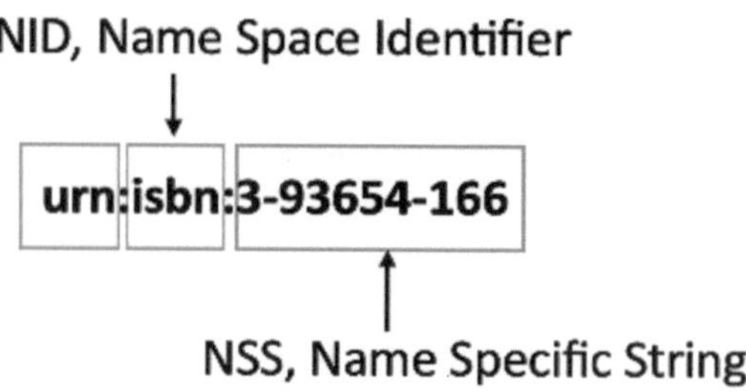

Fig. 3.12 Example of a URN

each NID. Since developing a custom name syntax can be laborious, when creating a URN it is often easier to request a unique NSS from an already existing NID group (represented by an administrative institution or similar). As far as character encoding is concerned, the URN standard is somewhat peculiar. The NID part may only consist of letters, numbers, and the "-" character. For the NSS string, the same character set as for URLs may be used. Non-displayable characters are encoded using escape sequences in UTF-8. The standard does not otherwise specify how to access metadata, nor does it go into detail about resolution mechanisms. It merely requires that resolution should be efficient.

Examples of URN implementations include:

- urn:ietf: Identifiers for IETF (standard) documents
- urn:ISSN: Integration of International Standard Serial Numbers into the URN concept
- urn:ISBN: Integration of International Standard Book Numbers into the URN concept
- urn:NBN: Identifiers for books (integration of the National Book Number)

In summary, the URN standard consists only of a set of requirements (both syntactic and conceptual), but it is left to the implementing party to decide what is ultimately made of it. The goals of URNs are ambitious and have been partially realized in many real-world systems. However, it should not be expected that a persistent identifier will automatically result from mere requirement specifications.

The most fully developed and elaborate persistent identifier system is the Digital Object Identifier System (DOI) [29]. In contrast to the URN, the DOI is based on a comprehensive and unified infrastructure. Technically, the DOI relies on a handle service for resolution, while a dictionary is used for metadata management. Although the DOI documentation explicitly states that these two components could in principle be replaced by something else, in practice the DOI is inseparably linked to them. The goals of the DOI system are as follows:

- Identification,
- Persistence,
- Resolution support,
- Access to metadata,

Fig. 3.13 Example of a DOI

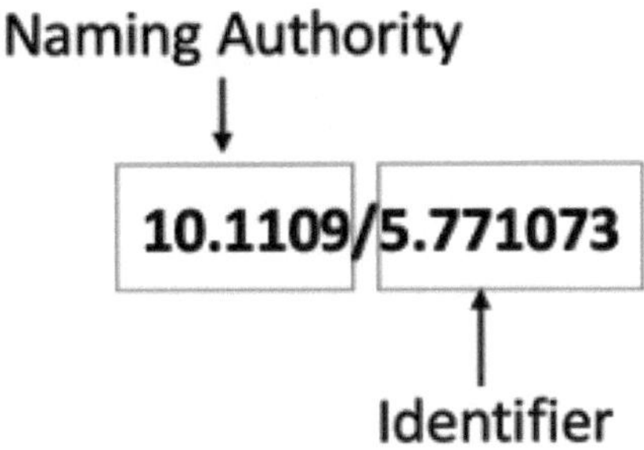

- Support for semantic interoperability between different (non-DOI) systems,
- Support for the metadata layer.

In addition to the relatively high degree of uniformity provided by the use of the handle system, an important principle of DOI is that it is not intended to create a new, standalone system, but rather to provide a means of integrating existing identifier systems. Two different options are offered for embedding conventional identifiers into a DOI: integrating them directly into the DOI syntax or instead including them in the metadata. The basic structure of a DOI is shown in Fig. 3.13.

The DOI consists first of a **prefix**, which designates the naming authority. This is followed by the **suffix** after the "/" character, which designates the existing identifier. Unicode is used for the character encoding of the entire name. The advantage of this approach is its universal applicability in almost all countries of the world. However, it should be noted that a DOI must first be encoded as a URL before being sent over the Internet. It is also worth noting that there are no restrictions regarding the length of the DOI. Both prefix and suffix are case-insensitive (similar to the first part of a URL). The entire DOI system is managed by the **International DOI Foundation (IDF)** [29]. The IDF is thus the highest authority, under which a number of Registration Agencies (RA) operate. An RA manages digital objects under a common concept, which is expressed through the Application Profile (AP). Some further properties of DOIs are that as little semantics as possible should be used in the identifier. Changes in ownership are also regulated by guidelines. The way links are redirected is also specified. Metadata handling is defined in the application profile of an RA. All RAs share the Kernel Declaration, which defines a minimum language scope for the description and structuring of metadata.

The **resolution** is, as in any identifier system, a main component of the DOI. To achieve its ambitious goals, the DOI relies on the handle system. The basic requirements for handles are as follows:

- Security
- Distributed administration
- Efficient resolution
- Support for multiple instances
- Support for multiple attributes

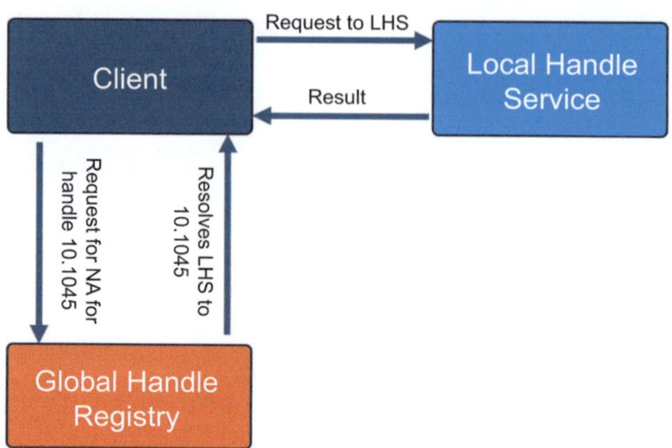

Fig. 3.14 An example handle system

The **Global Handle Registry (GHR)** serves as the highest administrative authority in the handle system. As shown in Fig. 3.14, it is fundamentally the first point of contact for all requests, as it provides the address of the corresponding Local Handle Service (LHS) for a given handle. The address of the LHS is also transmitted as a handle. In a second step, the client can then request the desired resource from the LHS.

Each LHS is associated with several service sites, which then contain the actual servers. A request to an LHS is sent in parallel to all service sites, which reduces response times and also makes the system fault-tolerant. Since both the number of service sites and the number of servers per service site can be varied, the whole system is also highly scalable. A handle is always resolved to a set of values. Each element of this set is assigned a data type, which itself can be retrieved as a handle in the system, the access rights, the administrator ID, and, of course, the data itself.

Thus, the DOI system fulfills most of the requirements set out at the beginning. It is flexible and efficient and is therefore, in principle, best suited for long-term storage. However, a clear disadvantage is that it is a commercial system managed by the IDF. Therefore, the lifespan of an identifier is likely tied to the lifespan of the DOI. In addition, DOI services are currently subject to fees. This is not the case with any of the other systems. The URN protocol has ambitious goals but is still in its infancy due to the lack of a larger number of reference implementations. ARKs offer an interesting approach without requiring much technical effort. In contrast, PURLs can only be considered as a transitional solution. Regular URLs have very poor persistence properties and should not actually be used in this sense at all. At present, however, they are the most widespread form of identifiers for digital objects, and this is unlikely to change any time soon.

Persistent identifiers and the associated infrastructures form the basis of content organization.

3.5   Content Organization

The provision of digital media is becoming increasingly important. In particular, internet-based access to multimedia has led to infrastructure service providers that distribute multimedia assets worldwide to guarantee optimal bandwidth and access for as many users as possible. Such **Content Delivery Networks (CDN)** [30] can now be booked as a service from common cloud providers such as Amazon or Microsoft; the respective provider then takes care of the distribution, management, possible conversion, and delivery of the content. Here, too, the concept of unique persistent identifiers is an important foundation to enable such services in the first place. Without a CDN, all users would have to obtain the content from a central server. In the event of a server failure, no content could be delivered, and all users would need sufficient bandwidth to access the central server from their region (see Fig. 3.15). By introducing a CDN, the content is distributed from the central server to additional servers worldwide (by copying, see Fig. 3.16). Thus, the content is available redundantly, and the failure of a server would only affect the respective region. In addition, users can use a server located in their region, which generally leads to significantly better access times.

Depending on configuration, CDNs can be optimized for the fastest possible delivery (performance) or for minimal traffic (bandwidth). To this end, CDNs offer

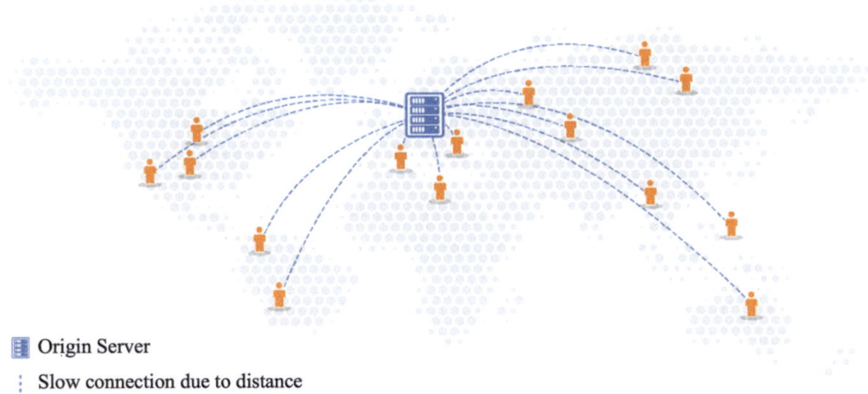

Fig. 3.15 Media distribution without CDN

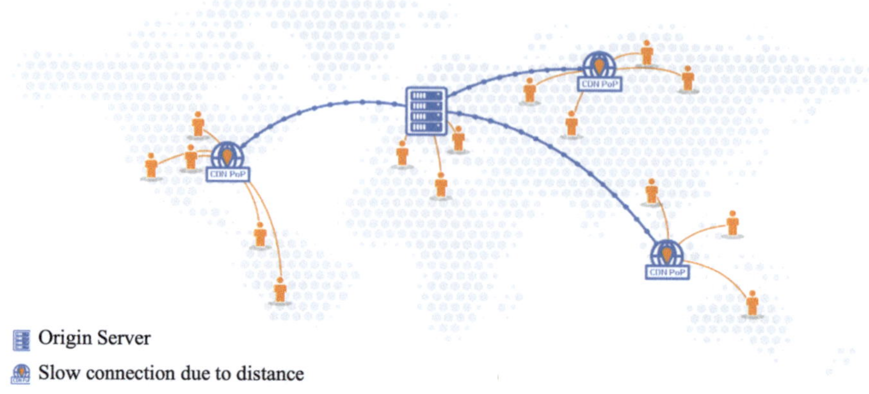

Fig. 3.16 Media distribution with CDN

the option to automatically store content in several formats of different sizes and deliver a specific format depending on the client. In the context of CDNs, the following roles and services are typically used [30–32]:

- **Content Provider:** The owner or provider of the content.
- **Authorization:** The content provider grants the CDN provider permission to deliver content.
- **Reporting:** The content provider requests a performance analysis from the CDN provider to evaluate the service quality of the CDN provider and gain access to other relevant data.
- **Source:** The content provider transmits a copy of the content. At this point, for example, digital signatures of the content or end-to-end encrypted content can also be specified. In many cases, distribution via CDN is transparent, i.e., the CDN provider has no knowledge of the actual content. The exception is any conversion to other formats.
- **Content:** Digital information created or licensed for distribution in the form of metadata. In most cases, this is a copy of the source material on individual CDN nodes.
- **Request:** The user requests the content provider to display or store data locally. The CDN provider selects the best location and the most suitable format of the content for this purpose.
- **Delivery:** The CDN sends the content from a specific node to the user.
- **User:** The user or technical application that wants to request data from the content provider. In many cases, this is the end user directly. However, CDNs are also often used internally within applications. In this case, the user is an internal, technical user within the application.
- **Delivery Node:** The main task is to deliver content to the user. Delivery nodes are servers with caches where content is provided. Content can be stored

manually on these nodes (push CDNs), or the delivery nodes can request content from the origin nodes (pull CDNs). The main advantage of push CDNs is that the content is available to users who retrieve it without delay. The main disadvantage is that the content provider must proactively "push" the content after each update. The difference from pull CDNs lies mainly in the automatic retrieval of content from the content provider. The main disadvantage is the initial speed of content delivery: when the user retrieves the content for the first time, delivery is just as fast as if the content provider were not using a CDN.

- **Storage Node:** The main goal is to store copies of the original data that are distributed to delivery nodes. A hierarchical model can be defined for storage nodes, allowing multi-level caching. Storage nodes are typically slower than delivery nodes but usually have more disk capacity and redundant backups of the content.
- **Origin Node:** These are the main content sources that allow the transmission of content over the network or the infrastructure of the content owner.
- **Control Node:** The main task of control nodes is to host management, routing, and monitoring components of a CDN. At the content level, there are essentially three different types:
- **Dynamic Content:** Content that is automatically generated using one of the common web programming languages such as PHP, Ruby, or Java.
- **Static Content:** Content that usually does not change very often and does not need to be recreated, such as images, CSS, and JavaScript.
- **Streaming Content:** Videos or audio files that are played back using a web browser control.

In addition to these general characteristics, it is up to the application architect to design further options based on CDNs. For example, CDNs can be designed and used for private use (i.e., use within a closed user or application group) or for public use (unrestricted access). There is also often the so-called multi-CDN approach, in which CDNs from several providers are combined to achieve additional fail-safety through redundant providers. The main advantages of a CDN are faster load times, lower bandwidth usage, better performance, and flexible scalability. The disadvantages are higher effort, the cost of the service, and a certain loss of control.

> CDNs are solutions for distributing multimedia.

Especially in the field of video and audio distribution, content is typically no longer downloaded by the user and then played locally. Instead, so-called streaming [30] is used, i.e., only the segment of a multimedia asset currently being viewed by the user is loaded and played directly. Local buffering and the associated complete prior download are no longer necessary. To enable streaming, the corresponding multimedia asset must be in a format that allows users basic navigation (play, pause, fast forward, and rewind). However, this can be problematic, especially with video formats, because video compression, for example through the introduction of P- and I-frames in the MPEG format, no longer allows direct access to individual frames, as only delta information is stored between these frames. In such cases, fast-forwarding or rewinding in the video does not access individual frames, but only the respective fully available keyframes. For most streaming applications, this is not a problem, but there are more and more use cases where collaborative multimedia platforms also require frame-accurate video or audio processing. For example, Adobe Premiere or Apple Final Cut offer collaborative video editing, or Apple Logic Pro offers collaborative audio editing based on streaming. Here, a low-resolution copy of the original material is often used on local computers to perform frame-accurate actions and keep track of multimedia projects. These actions are simultaneously applied in the background on the high-resolution material server-side by the provider of the collaboration platform and distributed to the other participants in the collaboration project. This is also a form of streaming.

In streaming, two types are generally distinguished:

- **On-Demand Streaming:** Here, data is transferred from the server over the network (or the CDN) to the client. Playback begins during transmission. Buffering is necessary to avoid bandwidth issues and ensure seamless playback. Fast-forwarding, rewinding, and pausing are possible. HTTP and FTP are usually used as transmission protocols.
- **Live Streaming:** Here, the content is provided in real time (e.g., for live events but also for TV broadcasters who provide their stream live on the internet). Typical protocols are RTP, RTCP, RTSP, SIP, or SRT. Fast-forwarding is only possible up to the current live position, and pausing and rewinding are usually only possible for a certain period of time.

According to a 2021 survey, 81% of all German internet users aged 16 and over watch videos via streaming at least occasionally. In 2019, it was 79%. In 2021, 38% use paid video streaming subscriptions—of these, 32% now use two subscriptions and 29% even three or more. 76% of internet users aged 16 and over in Germany listen to music via streaming services such as Spotify, TIDAL, Apple Podcasts, or Deezer at least occasionally, with almost half (48%) using paid services. This means that audio streaming has continued to gain popularity: in 2019, only 72% of internet users aged 16 and over in Germany listened to music via streaming services, two-thirds of them daily. In 2019, one in five not only used free offers but also paid for music streaming.

3.6   Distributed Information Retrieval

Under certain conditions, it is advantageous or necessary to search more than one data source at the same time, for example because no central index is available. In a company, for example, there may be no central index because the data sources

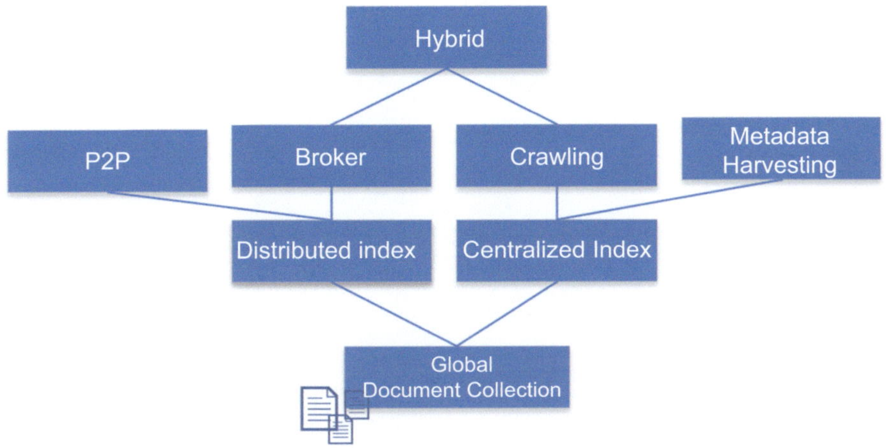

Fig. 3.17 Taxonomy of distributed IRS [35]

of individual departments should not be openly accessible, but only found and viewed under certain conditions. In such cases, a solution is to have the inventory available via the intranet processed only by department-specific IRS. The index of all company data is then only available to authorized persons. Another conceivable situation is one in which the index of a collection grows beyond a manageable size. Such situations lead to the requirement that not just one, but several IRS or indexes are necessary for automated access to a document collection. In this context, the term distributed search (or "Distributed Information Retrieval", DIR) is also often used.

This section is based in particular on the works of Callen in [33], Crestani in [36], and Croft in [34], which are also recommended for further reading and deeper study. To facilitate entry into the topic, the DIR taxonomy introduced by Crestani (see Fig. 3.17) is first presented. This provides an initial overview of the different challenges. The taxonomy starts with a global document collection, which, for the reasons mentioned above, can be split into a distributed index or exist centrally. A distributed index is searched using broker- or peer-to-peer-based (P2P) approaches, whereas for central indexes, crawling or metadata harvesting are possible approaches.

Let us first consider DIRs that operate on a distributed index. Afterwards, DIRs that operate on a central index will be briefly outlined: in a peer-to-peer-based DIR, it is assumed that each IRS and its associated index form a unit. All units are interconnected. If a query is submitted to unit A, it is forwarded to the connected units and an attempt is made to resolve it. All responses are returned to the sending unit after processing and merged by it. This procedure is schematically shown in Fig. 3.18.

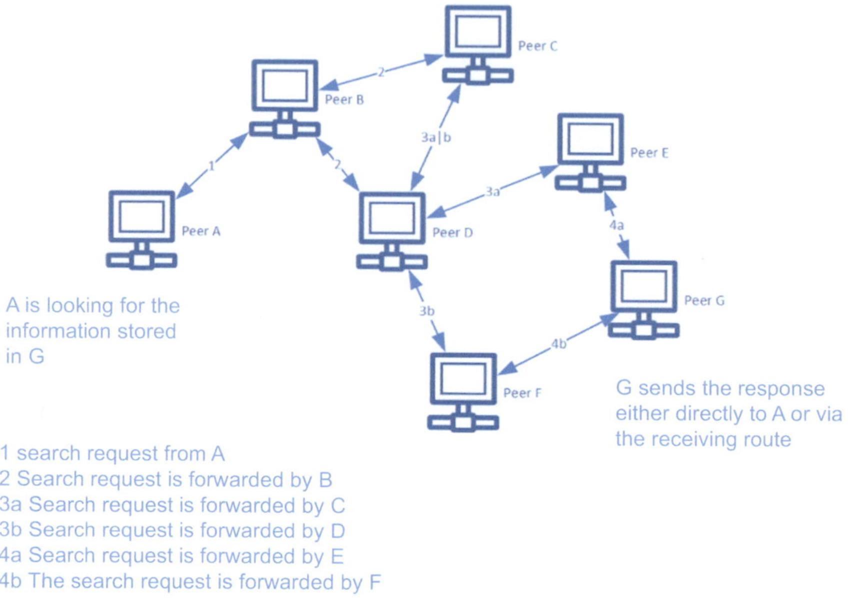

Fig. 3.18 Peer-to-peer network according to Crestani [36]

In broker-based architectures, communication with users takes place via a broker. The broker receives the query and forwards it to the connected IRS. The results calculated by the connected IRS are then returned to the broker, which performs the merging (see Fig. 3.19).

The broker-based architecture presented here is based on the mediator architecture model by Wiederhold [37], in which he describes what was then a new architectural concept in which so-called mediators act as intermediaries between data sources and applications. These mediators perform semantic transformations and integrations to make heterogeneous, distributed data sources usable. This shifts the burden of data processing from the end applications to specialized intermediate layers. Wiederhold argues that such architectures are necessary to cope with the increasing complexity and distribution of modern information systems, which can be clearly seen in the architectures presented here and was taken up by Crestani [36].

An application that is commonly addressed via a central index is metadata harvesting. It also works with a broker-based approach. An online and an offline processing phase are distinguished. In the offline phase, metadata is retrieved from the broker using appropriate protocols. In the online phase, users of the system can submit queries to the broker, which returns a set of relevant outputs based on the available metadata (see Fig. 3.20).

Each of the DIR variants presented here poses different challenges when implementation is attempted. In this subsection, broker-based approaches will therefore be examined in more detail as representative.

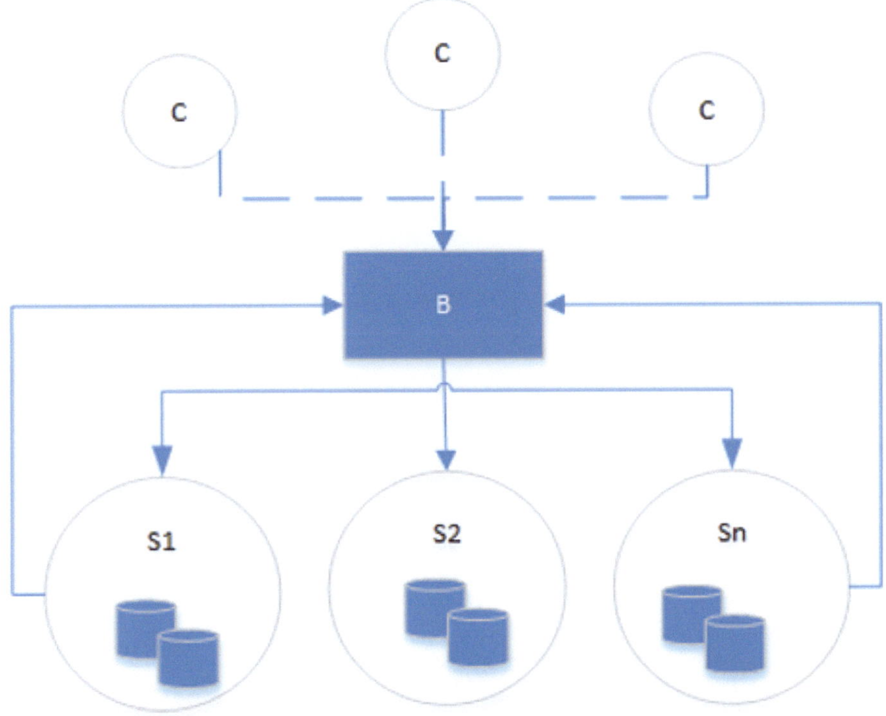

Fig. 3.19 Broker-based architecture according to Crestani [36]

Within broker-based approaches, there is a requirement to query different and distributed databases (or document collections). With regard to this requirement, three overarching challenges can be identified, which are first introduced in general. Subsequently, these overarching challenges form the structural framework for the further structure of this subsection.

- **Resource Description:** The contents of a collection must be described in such a way that only those queries are forwarded to this source that can actually be answered by it.
- **Resource Selection:** Based on a database description, it must be decided automatically which of the collections is suitable for the query.
- **Result Merging:** The set of results returned by each connected IRS must be merged for a consolidated output.

There are different architectures for distributed IR. However, all must be able to describe resources, select them, and merge results.

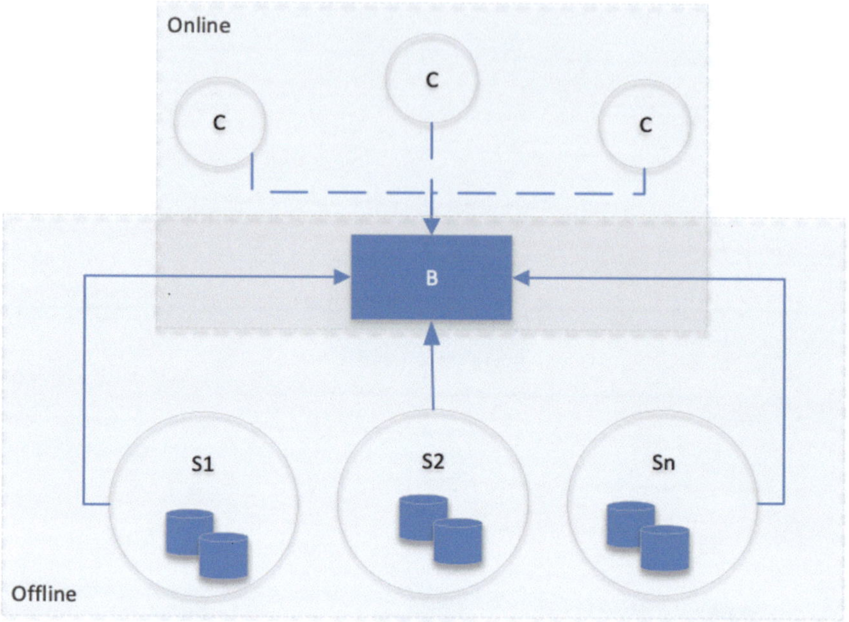

Fig. 3.20 Metadata harvesting according to Crestani [36]

3.6.1 Resource Description

Descriptions of document collections are needed to make effective and efficient decisions about which of the available sources should be used to answer a query. The challenge here lies on the one hand in selecting good features to describe the contents of a collection, and on the other hand in the automated generation of the description. The latter is particularly relevant in dynamically growing databases.

Descriptions can include different features of the documents in a collection. First, the content can be described using general features. In the case of full-text databases, these can be statistics (for example, term frequency), as are also maintained in an index, or content information such as representative terms (keywords). In the case of multimedia databases, these can be technical, content-based, or semantic feature representations. A list of possible features, divided into pre- and post-retrieval features, can be found in Kopliku [38], Tables 3.3 and 3.4. Furthermore, information on response times, costs incurred, or content overlaps between collections must also be considered.

When capturing features, a distinction is also made regarding the environment. Not all information is equally easy to access or query:

- **Cooperative environment:** An integrated source allows the broker to access all indexes and associated documents.

Table 3.3 Pre-retrieval features for describing databases

Pre-retrieval feature	Description
Vertical intent terms	Some search terms imply searching for specific components, such as images, videos, or music
Named-entity type	The query contains named entities and can be assigned to a source on this basis
Query length	Length of the terms in a query
Query logs	Repeated occurrence of a query
Recent popularity of the query	Number of times a query is repeated
Clickrate Analysis	Implicit feedback data
Source representation	Implicit feedback data
Category representation	Classification of the query into a higher-level class (e.g., sports or cooking)
Evidence for navigational need	Navigational needs

Table 3.4 Post-retrieval features for describing databases

Post-retrieval feature	Description
Vertical relevance score	The RSV of an IRS on the source in question
Number of results	Number of results
Freshness of documents	Recency of the information
Contextual score of results	Consideration of contextual factors in the broadest sense
Geographical-context score of results	Geographical context

- **Non-cooperative environment:** An integrated source only answers queries and does not allow access to the data inventory or associated indexes themselves. This is the case, for example, when documents from individual departments of large companies are to be searched only internally.

While in cooperative environments information for generating descriptions can be easily obtained, the situation is much more difficult in non-cooperative environments. Here, brokers can only attempt to approximate a description. However, this always involves a certain degree of uncertainty.

A simple solution for querying information in non-cooperative environments is so-called query-based sampling [39]. This provides a way to generate an approximation. In these methods, queries are submitted to the system and the result set is examined in order to generalize to all documents contained in the collection. Implementations of this approach include submitting test queries in which word dependencies are tested and documented by checking the co-occurrence of two words in the query against occurrences in the data inventory. The challenge with this approach lies particularly in selecting good queries and determining the number of queries that must be submitted to the system to obtain a meaningful description.

3.6.2 Resource Selection

The selection of the collection to be considered for answering queries is carried out based on the so-called database description. The fundamental requirement is to determine which of the connected IRS is most likely to return the best results for the query. Thus, the requirements here are reduced to a relevance ranking problem of resources.

As an example, the method by Callen is introduced here. He bases his approach on the assumption that the selection of a data resource should also satisfy an information need. The advantage of this is that a query can be used both for the selection of a resource and for a document. The applied method is based on a so-called Bayesian Inference Network (see Fig. 3.21).

In this representation, each resource $(R_1 - R_j)$ is represented by a set of nodes (as representations of index terms) $(r_1 - r_k)$. In contrast, the information need is expressed in the form of a query q. The query itself again consists of a set of nodes (as representations of the terms in the query). The probability with which a query describes a resource is, as usual, denoted by $P(q|R_i)$. However, the network again works with representations of documents and the query. Callen calculates the probability $P(r_k|c_i)$ based on a variation of TF/IDF:

$$T = \frac{df}{df \cdot 50 + 150 \cdot cw/avg - cw} \tag{3.14}$$

$$l = \frac{\log(\frac{C+0.5}{cf})}{\log(C + 1.0)} \tag{3.15}$$

$$P(r_k|c_i) = b + (1 - b) \cdot T \cdot l \tag{3.16}$$

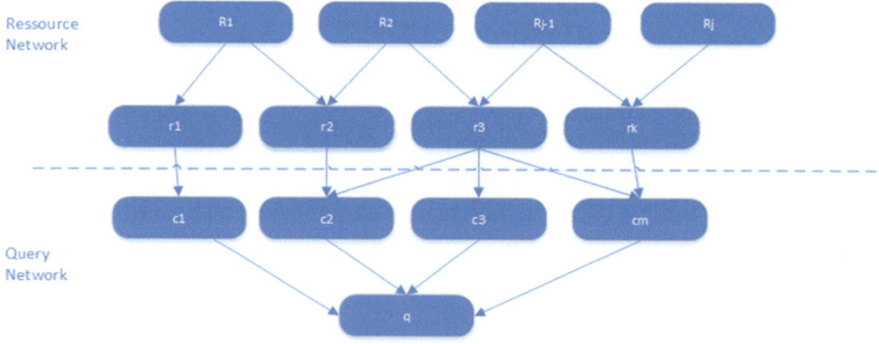

Fig. 3.21 Bayesian inference network

The following applies:

- df: Number of documents in R_i that contain rk
- cw: Number of index terms in the resource R_i
- $avg - cw$: average number of index terms in each resource
- C: Number of resources
- cf: Number of resources that contain the term r_k
- b: minimum "Belief Component" (usually approx. 0.4)

The values $P(r_j|R_i)$ can be combined using different operators, e.g., by simple summation and normalization or also by taking weights into account. In this way, it is possible to statistically determine those resources that are most likely to be tasked with processing the query.

3.6.3   Result Merging

After a set of results for a query has been identified, the final step is to merge the result sets from the individual IRS. Various problems arise in this process. First, it cannot be assumed that the broker knows how the RSV for a document to a query was calculated by an IRS, or whether the calculations of different connected systems are compatible with each other. In addition, in many cases it cannot be ruled out that there are document overlaps (the same documents are managed by more than one IRS) among the connected IRS. To address these challenges, a number of possible solution approaches are described below. First, approaches for normalizing relevance measures are considered. Subsequently, methods are introduced that can handle document overlaps. The following sections are based on Wu [40].

First, a simple approach will be discussed that can be used to normalize RSVs from a set of connected IRS $L_i(i = 1, 2, \ldots, n)$ in a cooperative environment. A simple applicable method for normalizing scores in a cooperative environment is to normalize them into the interval $[0, 1]$. The following function accomplishes this requirement:

$$s_{ij} = \frac{r_{ij} - \min -r_i}{\max -r_i - \min -R_i} \tag{3.17}$$

Here, $\min -r_i$ corresponds to the minimum RSV of all RSVs from the set L and $\max -r_i$ to the maximum RSV from the set L. The variable r_{ij} stands for the RSV to be normalized.

The function s_{ij} is also called the zero-one linear function and is optimal when the connected IRS consistently deliver good results. Consistently delivering good results here means that relevant documents for a search have been assigned particularly high RSVs—and non-relevant documents particularly low RSVs. If an optimal RSV calculation is present, then the zero-one linear function can provide a good distribution of RSVs into the normalized interval. If the evaluation of the IRS under consideration is not as good, it may be desirable to restrict the results to a narrower interval $[a, b]$, where $(0 < a < b < 1)$. The approach in the latter case is also referred to as the fitting model. The challenge in the fitting model is to find the best values for a and b. However, this is not a trivial process with a clear procedure. One suggestion Wu makes is to use linear regression. Without going further into the calculation of a linear regression, it should be briefly noted that this method is generally used to calculate relationships between variables.

Merging result sets in uncooperative environments is significantly more complex. In such cases, it may not be known how RSVs were calculated by an IRS; in some cases, there may not even be RSVs at all, but only a ranking of the results. An example of this is the Reciprocal Rank Method (RRM). This method simply compares the positions of result sets that only have a ranking and assigns documents a new or own RSV based on this. The Borda Count method works differently. It is based on the idea of a democratic voting procedure. In this method, elements from m result lists are assigned descending values. For a result set of n documents, the highest-ranked document would receive the value n, the next one $n - 1$, and so on. The sum of the individual results yields the fused relevance score.

3.6.4 Aggregated Search

So-called aggregated search is very similar to distributed IR as described by DIRs. In particular, aggregated search also includes the already introduced phases of database description, database selection, and result merging. The main difference from DIRs lies in the requirements for the result set that users expect from an aggregated search. The goal of aggregated search is usually not to output complete documents as information carriers, but rather to extract only the required (sought) information from different documents and deliver it to users in a meaningfully combined (aggregated) form. The motivations for aggregated search can be summarized as follows (see also Kopliku [38]):

- Distributed information: The sought information is not contained in a single document, but is instead distributed across a large number of documents.
- Non-focused results: Users are often interested in only a small part of a found document. Instead of an entire document, only the sought information should be delivered to the users.
- Ambiguity: Queries can be ambiguous. For example, the question about Java can refer to an island or a programming language. Outputting a variety of possible facets is one way to provide users with a tool to disambiguate ambiguous queries.

In summary, Kopliku defines aggregated search as follows: "In its broadest definition, we say that aggregated search aims at retrieving and assembling information nuggets, where an information nugget can be of whatever format (text,

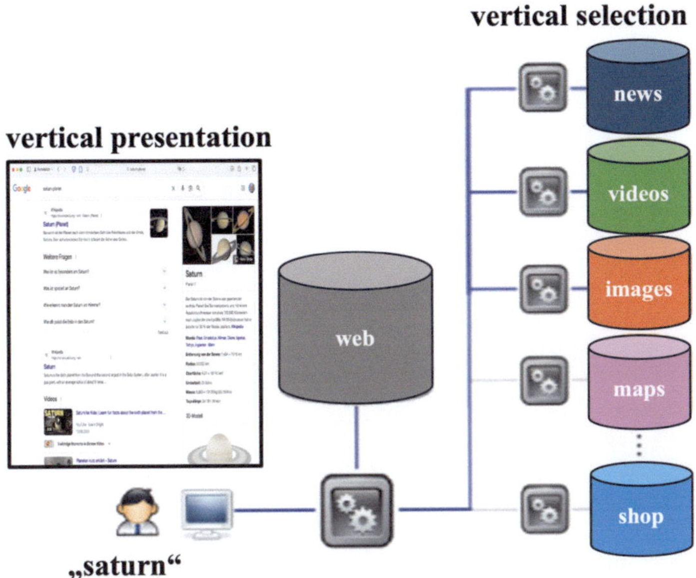

Fig. 3.22 General procedure of an aggregated search

image, video, etc.) and granularity (document, passage, word, etc.)." According to Kopliku, an information nugget is: "a generalization of content of some granularity and multimedia format."

Aggregated searches are mostly found in the context of web applications that create a comprehensive picture of a sought-after context from a set of highly specialized IRS (see Fig. 3.22). The components of an aggregation are referred to as verticals.

In aggregated search, three consecutive phases are generally distinguished (see Fig. 3.23): Query Dispatching, Nugget Retrieval, and Result Aggregation. These will be presented in detail in the following subsections.

3.6.4.1 Query Dispatching

The query processing phase (English: "Query Dispatching") includes all actions related to capturing the user's information need and preparing it for forwarding to the connected document collections or IRS. In particular, three overarching requirements can be distinguished:

- **Source selection:** Source selection describes the process of identifying an ideal subset of all available IRS. The selection process, analogous to the first two phases of implementing a broker-based DIR (resource description and selection), involves identifying, among a set of available sources, those that can best meet the users' information needs. This step is especially necessary when a large number of possible sources are connected and long processing times are

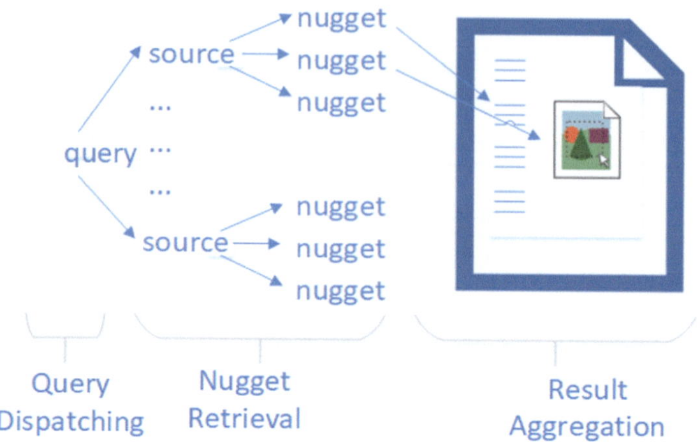

Fig. 3.23 Framework of an aggregated search

imminent. Solutions for implementing a suitable source selection procedure include, for example, analyzing the query for certain keywords, such as weather or news. These keywords can help directly identify a suitable source, as long as the source has been appropriately registered. Recognizing specific entities (English: "Named Entity Recognition," NER) can also be helpful in achieving better automation in source selection. If a query is recognized as asking about the person Oscar Wilde, sources with relevant content can be queried directly. Additionally, further relevant queries can be formulated that search for related persons (e.g., other authors of that era).

- **Query intent:** A requirement related to source selection is to automatically recognize the intent of a question (English: "Question Intent"). The intent can be distinguished, for example, with respect to searching for certain types such as: why, where, when, or yes-no questions.
- **Query reformulation:** Depending on the selected sources, adjustments can be made or may even be necessary. This is the case, for example, when more than one source is queried via a common (global) interface. In this case, the user's query may need to be adapted to the common interface of a connected source.

> Query Dispatching selects the appropriate information sources in a distributed architecture and adapts the query accordingly.

3.6.4.2 Nugget Retrieval

An "information nugget" in the context of nugget retrieval refers to a specific unit of multimedia information that is to be processed and found. In principle, a variety of methods can be used for this requirement, which cannot be discussed in detail in this chapter but will at least be mentioned.

In the previous subsections, documents were described as information carriers, with documents in this function being seen as a kind of information container. These information containers contain the information sought via an IRS, but are usually not the information sought itself. So-called focused retrieval methods take this fact into account and implement the search for the aforementioned information nuggets within a document. These methods can include, among others, passage retrieval, XML retrieval, and entity searching and ranking.

Passage retrieval methods are those that implement a search for document passages of variable length, as is used, for example, in question-answering systems to identify relevant passages from texts. Retrieval methods in semi-structured data, such as XML retrieval, use not only the textual content of a document but also the information conveyed by its structural elements. In XML retrieval, two query forms can be distinguished, depending on whether the XML files are more content-oriented ("Content-only," CO) or structure-oriented ("Content and Structure," CAS).

Analogous methods can be established for all multimedia objects. For example, not the entire video is returned, but only the segment/scene that matches the query. For images, it may be that only certain sections are returned, or at least a marking in the image highlights the relevant section. All of this, of course, depends on the quality and type of features that could be captured during the indexing process. If, for example, only the title is available as a feature for a video, the returned "nugget" would be identical to the entire video. However, if scene detection or annotation (e.g., with MPEG-7) has been performed, then exactly those elements from the video can be returned as nuggets that answer the query.

> Nuggets are units of information that describe the results of distributed queries.

3.6.4.3 Result Aggregation

The aggregation of the found information nuggets is the final phase. Here, the found information is transferred into an aggregated state ("Result Aggregation"). The aggregation process includes various distinguishable actions, which are described below:

- **Grouping:** To facilitate the presentation of information, individual information nuggets can be combined using classification methods. This corresponds to grouping information nuggets according to common criteria (e.g., event locations when searching for Oscar ceremony events). Classification methods can also be used for this application.
- **Sorting:** Nugget retrieval can result in a large number of nuggets. For easier further processing, this set must also be sorted. Sorting can be performed based on common features. Possible approaches include using the RSV of focused IR methods as well as time or location references.

- **Merging:** Unlike grouping, where information is bundled according to certain relationships, merging combines information nuggets into a new unit that can serve as the output of the process.
- **Splitting:** The reverse action to merging is splitting information nuggets into smaller components, for example, dividing a text into its paragraphs or sentences.
- **Extraction:** Extraction refers to the process of identifying stand-alone information nuggets. These can be, for example, images, videos, or named entities.

In general, aggregated search initially only includes methods that search for specific information from distributed sources and deliver it to users. The expressiveness of the results can be increased by adding explicit connections between the information units. This allows questions of relatedness between information units to be made directly visible. Such connections can be very general in nature (e.g., a jaguar "is a" creature) or very specific (e.g., Mozart "born in" Salzburg). What these connections have in common is that they can be expressed in the form of a triple.

Connections of a general nature can, for example, be derived via similarity calculations. The hypothesis when applying such methods is that if two results have almost the same RSV for a query, then it can be stated that both results are also very similar. Based on this consideration, "similar to" connections can be established. Conversely, if RSVs are very different, a "different to" connection can be derived.

> The results collected in a distributed manner are merged and sorted during result aggregation.

3.6.5 Parallel Processing of Queries

The preceding subsections mainly dealt with searching in distributed sources and aggregated search. The focus of this chapter is dedicated to the challenges that arise when the index of an IRS exceeds a certain size and measures must be taken for its efficient creation and querying. The content of this chapter is based in particular on the book by Büttcher [41].

Indexing and querying very large data collections is a challenge in terms of memory consumption and runtime behavior of an IRS. Prominent solutions to address both problems are partitioning the index into partitions (partitioning) or replicating the index across multiple servers (or nodes of a network). Partitioning the index is also referred to as intra-query parallelism, and replication as inter-query parallelism:

- **Intra-query parallelism:** Each node in a network holds a significantly smaller partition of the overall index, which can be processed quickly due to its size.
- **Inter-query parallelism:** Each node in a network holds a replica of the index.

	Anthony and Cleopatra	Julius Caesar	The Tempest	Hamlet	Othello	Macbeth
Anthony	1	1	0	0	0	1
Brutus	1	1	0	1	0	0
Caesar	1	1	0	1	1	1
Calpurna	0	1	0	0	0	0
Cleopatra	1	0	0	0	0	0
mercy	1	0	1	1	1	1
worser	1	0	1	1	1	0

Fig. 3.24 Partitioning by terms or documents

A set of queries can thus be processed in parallel on the available nodes. The latter is generally rather simple and will therefore not be the focus of this chapter. Instead, two methods of intra-query parallelism will be discussed: document-based partitioning and term-based partitioning.

- **Document-based partitioning:** Each node in the network holds an index for a subset of all documents.
- **Term-based partitioning:** Each node of the index holds an index for a specific set of all occurring terms.

Both approaches are also illustrated graphically in Fig. 3.24. The first row of the table denotes all documents of an example collection, and the first column all terms occurring in the collection. The red borders simulate partitioning by documents, and the blue borders partitioning by terms. The following subsections examine the challenges and approaches of both partitioning methods.

3.6.5.1 Document-Based Partitioning

The general approach in document-based partitioning is to split the collection into a set of partitions and have each resulting partition indexed by a node in the network (red frames in Fig. 3.24). Due to this property, the resulting index can also be referred to as a local index. The partitions created on the connected nodes are managed by a master node (English: "Receptionist"). All queries are sent to the Receptionist, who then distributes the queries to the connected nodes. The nodes process the queries and send back the top k results. The responses from the

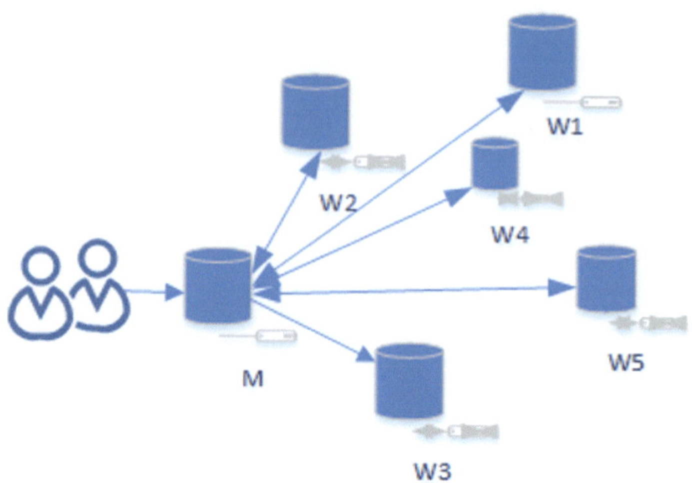

Fig. 3.25 Query processing in document-based partitioning

connected nodes must then be merged and output by the Receptionist. This procedure is illustrated in Fig. 3.25.

In principle, there are three main challenges that should be briefly addressed.

- **Effective partitioning of documents:** How can the documents be optimally distributed across the nodes? Various considerations can be made here. One possibility is to group documents on a particular topic together. This can, under certain circumstances, result in good data compression. However, if a topic becomes very popular, the load on the corresponding server will be very high. Since similar phenomena can also occur with other grouping methods, Büttcher et al. suggest not considering relatedness, but instead performing a random distribution.
- **Effective response size:** If the number of relevant documents output by the Receptionist to the users is limited, the question arises as to how large the result set from each node may be that is merged by the Receptionist. The problem is that it is unlikely that all relevant documents will be returned by a single node. It is also unlikely that all nodes will return the same number of relevant documents, which leads to high network load. To address this problem effectively, various methods have been proposed that are based on probability calculations. These will not be discussed further here.
- **Effective processing:** Especially in the area of multimedia collections, it can also be helpful to split documents based on their multimedia content type (i.e., file format). For example, since video processing often requires special hardware, it makes sense to store all videos on machines with the appropriate hardware, while images or text documents can be stored elsewhere.

In general, the goal of document-based partitioning is, on the one hand, to store partitions in fast-access primary memory, and on the other hand, to keep partitions redundant (and thus fault-tolerant). In this setup, a set of servers is usually provided for performing the actual indexing and access, as well as a server that meets the requirement of distributing the tasks (indexing). The advantage of this approach is the parallel execution of processing steps, for which the following independence considerations must be made:

- The analysis (e.g., tokenization) of one document is independent of the analysis of another document.
- Sorting word frequencies can be performed independently.

An advantage of document-based partitioning is that even if a server fails, at least partial results can still be returned. This is because the participating nodes can operate independently of each other. This also results in less inter-server communication.

3.6.5.2 Term-Based Partitioning

Term-based partitioning distributes the entries of the index across a set of nodes in a network. Here too, the implementation of the approach assumes the existence of a Receptionist node. This node receives the query $q = \{t_1, t_2, \ldots, t_{|q|}\}$ and splits it into a set of k subqueries $\{q'_1, q'_2, \ldots q'_k\}$. The size of k is determined by the number of sets of terms in one of the contained subqueries. It holds that $k \leq |q|$. The Receptionist node distributes the resulting subqueries to the nodes that contain the elements of the subqueries. The terms that can be processed by a node are then added to an accumulating list of documents either with an AND or OR connection (Boolean query). This means that the matching documents of an index are only added to the accumulating list in the case of an AND connection if all terms of the subquery are present. In the case of an OR connection, it is sufficient if one term is present. Once all matching documents have been selected in this way, a global RSV is calculated in the Receptionist. The Receptionist then outputs the top k documents. The procedure is visualized in Fig. 3.26.

The advantage of term-based indexing is that the query only needs to be processed on the servers that contain the terms. The disadvantage is that if a server fails and is not redundantly available, the query cannot be answered. Scaling is problematic with this approach, since certain terms can occur in combination in very many documents and in many queries simultaneously. Some servers therefore require a lot of memory and have a high processing load. In some cases, this can make the index very large, and strategies should be considered to split or replicate overly large postings so that they can be processed in parallel on multiple servers. Another disadvantage is the increased communication overhead, which results from the fact that the participating servers must exchange ever-growing amounts of data. This is especially the case when using OR connections.

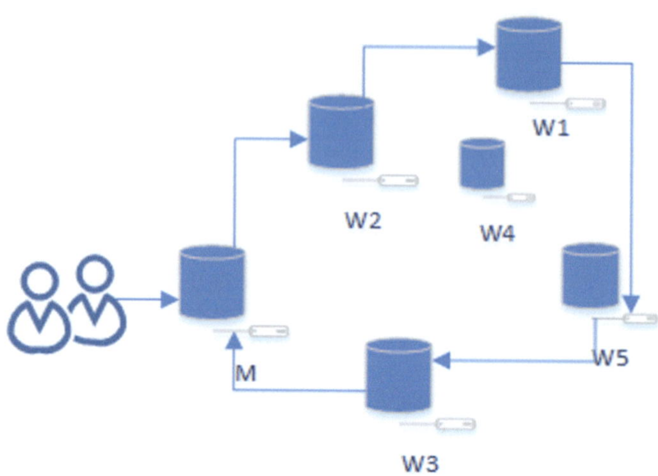

Fig. 3.26 Query processing in term-based partitioning

3.6.5.3 Fault Tolerance and Redundancy

Considerations for distributing an index become particularly relevant when very large data collections are to be processed. In addition to distributing the load for efficient processing, redundancy and fault tolerance are important topics in the context of IRS for large data collections with frequent access.

To optimally distribute the required computing resources, many computers can be used to process a query. However, the more computers are in the network, the higher the risk of a computer failure. If a certain fault tolerance has already been taken into account in the computation (as is inherently the case with global indexing), the failure of a computer has only a minor impact. In fact, the more computers are in the network, the less problematic such a failure would be. If this is not the case, the failure of a computer can lead to massive calculation failures. An example comparison will illustrate the impact.

Suppose a term-partitioned, replication-free index is distributed across 32 servers and one server fails. In this case, the failure is manageable, as only those queries whose terms are on the affected computer can no longer be answered. All other queries are unaffected by the failure. The probability that a randomly selected query consisting of three terms can be processed by the remaining 31 servers is:

$$(\frac{31}{32})^3 \approx 90.9\,\%$$

In the case of a document-partitioned index, however, the query can still be processed, but some results may not be included. The probability that j out of the top k results of a randomly selected query are not found is:

$$(\frac{k}{j}) \cdot (\frac{1}{32})^j \cdot (\frac{31}{32})^{k-j}$$

The probability that the top $k = 10$ remain unaffected by the error is thus:

$$(\frac{31}{32})^{10} \approx 72.8\,\%$$

The calculation clearly shows that a local index is significantly more susceptible to failures of the participating servers. To mitigate failures, three strategies in particular have become established:

- **Replication:** The indexes are stored multiple times (replicated, r). In the event that a server fails, the remaining servers can compensate for the failure. The advantage of this method is its simplicity and that it increases both throughput and fault tolerance. The disadvantage is that if the system is operating at its resource limits and r is chosen to be small, the remaining $r - 1$ replicas can quickly become overloaded.
- **Partial replication:** The effort associated with complete replication can also be reduced if replication is limited to important documents in the collection. The difficulty with this approach is making a reliable prediction as to which documents these are.
- **Dormant replication:** The indexes distributed across n servers can be split and stored dormant (i.e., on disk). This would mean that for n servers, each of the contained indexes vi can be split into $n - 1$ fragments and distributed. In the event that a server fails, the corresponding fragment can be retrieved from disk and distributed to the $n - 1$ remaining servers for use. Dormant replication has the disadvantage that it roughly doubles the storage overhead, since each of the n nodes must store $n - 1$ additional index fragments.

3.6.6   Distributed Indexing of Large Data Collections

The distribution of processing steps in IR to efficiently process very large collections is not limited to query processing. When creating the index for large data collections, it also makes sense to consider a parallelized approach. For this challenge, the so-called MapReduce method for efficient indexing will now be introduced. The advantage of MapReduce is that sets of data can be split across distributed servers for further processing. The difficulty lies particularly in the meaningful partitioning of the data to be processed. Applied to the task of distributed indexing in IR, we speak of JobTracker servers, which handle coordination, and TaskTrackers, which perform the partial computations. The following explanations of MapReduce are based mainly on Manning [3], but also on Croft [34].

First, it should be noted that the method is used in a wide variety of problems for processing large data collections. In principle, the method can be divided into two phases. In the first phase, simply put, a list of objects is transformed into another list of objects of the same length (map). In the second phase, a function is applied to this list that reduces the elements of the list to a single object (reduce). The general procedure is shown in Fig. 3.27.

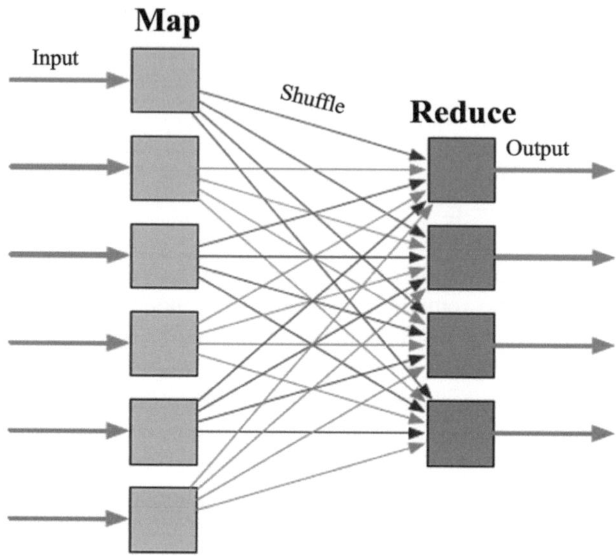

Fig. 3.27 General procedure of MapReduce

Assuming that the input is a set of string values (such as the textual representation of MM features), Map first transforms these records into a key/value pair.
These newly created pairs are then shuffled so that pairs with the same key are
grouped together. These should be grouped for further processing so that they can
be processed on the same server in the next step. Finally, in the reduce phase, they
are processed simultaneously and mapped to a value that is output. In summary,
this means:

- **Split:** The input is a set of data. The input data is split into meaningful parts.
- **Map:** The split input data is represented as key/value tuples ($K1$, $V1$).
- **Shuffle:** The transfer of data from the map process to the reducer is called shuf-
  fle. In the example in Fig. 3.28, the shuffle would, for example, transfer the
  tuple ($ceasar$, $(1, 1)$) or ($noble$, (1)).
- **Reduce:** The input for reduce is the key/value tuples mapped in the map phase.
  These are grouped into smaller sets of tuples ($K1$, $List(V1)$) and output. In the
  example, this would be a summary of the tuple ($ceasar$, $(1, 1))to(ceasar, 2)$.

For efficient indexing of large collections, this approach is now presented for
term-based inverted index partitioning based on the MapReduce method. As
an example, the approach according to Mannings et al. is introduced here (see
Fig. 3.29). In the map phase, the JobTracker is responsible for splitting the input
data and passing the split to a set of parsers. The parsers read successive parts of
a document and generate ($term$, $docID$) pairs. The pairs are written into j different

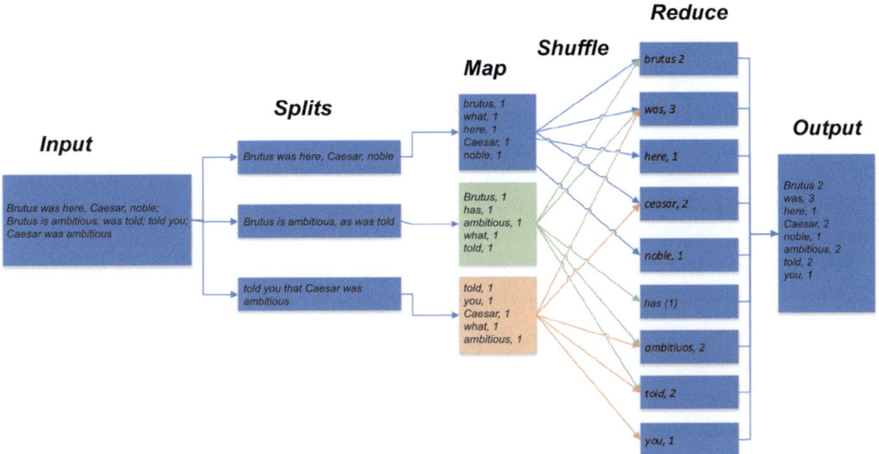

Fig. 3.28 Simple word count example with MapReduce

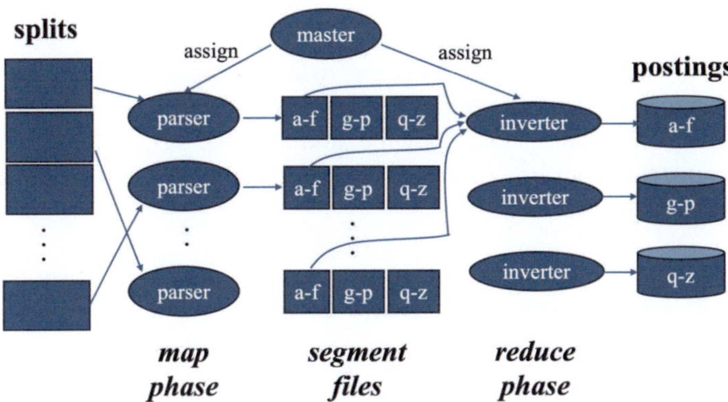

Fig. 3.29 Example of distributed indexing with MapReduce

partitions, each for a letter range (e.g., a–f or q–z). In the reduce phase, so-called inverters are used, which collect all (*term*, *docID*) pairs for a specific partition, sort them, and store them as postings in an inverted list.

3.7   Summary

This chapter systematically built up the fundamental understanding of Information Retrieval (IR). Starting from the definition and objectives of IR, the entire process of a retrieval operation was explained in a structured manner—from the formulation of an information need, through the indexing and processing of documents, to the output of relevant results. In particular, the conceptual model according to Fuhr was introduced, which provides a formal foundation for the technical implementation of IR systems.

Two central retrieval tasks were distinguished: **ad-hoc retrieval**, which responds to spontaneous user queries, and **filtering**, which continuously examines new documents for a fixed information need. Both tasks form the foundation of modern IR and MMIR systems.

Building on this, various IR model classes were introduced: **exact-match models**, which require a strict match between query and document (e.g., Boolean model), and **best-match models**, which interpret relevance as a gradual measure. The latter included the vector space model, fuzzy retrieval, the P-norm model, and probabilistic approaches such as the Binary Independence Model (BIM). This diversity of models makes it possible to adapt IR systems to different data structures, user needs, and uncertainties in query formulation.

Another focus was on **relevance feedback**, which serves as a method for the iterative optimization of search processes. The reader learned about explicit, implicit, and pseudo-relevant feedback strategies, which can be used to improve IR systems through user interaction or automated evaluation.

Finally, the chapter showed that IR systems can be understood as central components of modern information systems, which enable effective and flexible information retrieval through appropriate modeling, user centricity, and semantic representation. The reader should now be able to name basic IR concepts, understand their interrelationships, and classify various IR models in terms of their functionality and applicability.

3.8   Self-Assessment Questions

3.1—General Tasks

What are the core tasks of MMIR systems?

3.2—General Model

What does the RSV value denote?

3.3—General Model

What are the three conceptual units in the general information model?

3.4 -Exact-Match Model

State the result of the following queries based on the example document collection (see Table 3.1):

- $t_1 \wedge t_3$
- $t_1 \vee t_3$
- $t_1 \wedge t_2 \vee t_4$
- $t_1 \neg t_2 \wedge t_3$
- $t_1 \vee (t_2 \wedge t_4)$

3.5—Best-Match Models

What characterizes the vector space model?

3.6—Relevance Feedback

Explain the difference between explicit and implicit relevance feedback.

3.7—Relevance Feedback

Name the five most important strategies for the click-rate method.

3.8—Relevance Feedback

Describe the "Last Click > Skip Previous" strategy.

3.9—Relevance Feedback

What other possibilities for relevance feedback do you know?

3.10—Query Optimization

What does term co-occurrence mean?

3.11—Persistent Identifiers

Name 5 requirements for persistent identifiers.

3.12—Content Delivery

What role does the delivery node play in content delivery networks?

3.13—Distributed IR

What distinguishes cooperative from non-cooperative environments?

3.14—Distributed IR

What is a nugget?

3.15—Distributed IR

What phases does the transfer process of result aggregation have?

3.16—Distributed IR

What partitioning options for a collection do you know in distributed IR?

References

1. S. Büttcher, C. L. A. Clarke, and G. V. Cormack, *Information Retrieval: Implementing and Evaluating Search Engines*. Cambridge, Massachusetts: The MIT Press, 2016.
2. K. Aberer, *The Semantic Web*. Berlin, Heidelberg, New York: Springer Verlag, 2007, ISBN: 978-3-540-76297-3.
3. C. D. Manning, P. Raghavan, and H. Schütze, *Introduction to Information Retrieval*. Cambridge University Press, 2008.
4. The Text Retrieval Conference (TREC). "Datasets." (Jan. 2020), [Online]. Available: https://trec.nist.gov/data.html, Download: 11.10.2021.
5. R. Baeza-Yates, B. Ribeiro-Neto, and N. Fuhr, *Information Retrieval: Implementing and Evaluating Search Engines*. MIT Press, 2011.
6. J. Smith, *Knowledge Management: Concepts and Best Practices*. XYZ Publishing, 2020.
7. A. Kemper and A. Eickler, *Datenbanksysteme: eine Einführung*, 4., überarb. und erw. Aufl. München Wien: Oldenbourg, 2001, 608 pp., ISBN: 978-3-486-25706-9.
8. M. K. Buckland, *Information and Society*. Cambridge, MA: The MIT Press, 2017.
9. G. Boole, *An Investigation of the Laws of Thought, on Which are Founded the Mathematical Theories of Logic and Probabilities*. Macmillan and Co., 1854.
10. L. A. Zadeh, "Fuzzy sets," *Information and Control*, vol. 8, no. 3, pp. 338–353, 1965.
11. G. Salton, A. Wong, and C. S. Yang, "A vector space model for information retrieval," *Communications of the ACM*, vol. 18, no. 11, pp. 613–620, 1975.
12. G. Salton, E. A. Fox, and H. Wu, "Extended boolean information retrieval," *Communications of the ACM*, vol. 26, no. 11, pp. 1022—1036, Nov. 1983, ISSN: 0001-0782, 1557-7317. DOI: https://doi.org/10.1145/182.358466. [Online]. Available: https://dl.acm.org/doi/10.1145/182.358466 (visited on 02/15/2023).
13. F. Song and W. B. Croft, "A general language model for information retrieval," in *Proceedings of the Eighth International Conference on Information and Knowledge Management*, ser. CIKM'99, Kansas City, Missouri, USA: Association for Computing Machinery, 1999, 316–321, ISBN: 1581131461. DOI: https://doi.org/10.1145/319950.320022. [Online]. Available: https://doi.org/10.1145/319950.320022.
14. H. Zhang and Z. Su, "Relevance feedback in cbir," Jan. 2002, pp. 21–35, ISBN: 978-1-4757-6935-7. DOI: https://doi.org/10.1007/978-0-387-35592-4_3.
15. G. Salton and M. J. McGill, *Relevance Feedback in Information Retrieval*. Butterworth-Heinemann, 1986.
16. G. Salton and C. Buckley, "The smart retrieval system: Experiments in automatic document processing," in *Proceedings of the 1971 ACM SIGIR Conference on Research and Development in Information Retrieval*, ACM, 1971, pp. 107–117.
17. R. White, J. Jose, and I. Ruthven, "An implicit feedback approach for interactive information retrieval," *Information Processing and Management*, vol. 42, pp. 166–190, 2006. DOI: https://doi.org/10.1016/j.ipm.2004.08.010.

18. T. Joachims, "Optimizing search engines using clickthrough data," in *Proceedings of the Eighth ACM SIGKDD International Conference on Knowledge Discovery and Data Mining*, ser. KDD'02, Edmonton, Alberta, Canada: Association for Computing Machinery, 2002, 133–142, ISBN: 158113567X. DOI: https://doi.org/10.1145/775047.775067. [Online]. Available: https://doi.org/10.1145/775047.775067.

19. F. Radlinski and T. Joachims, "Query chains: Learning to rank from implicit feedback," in *Proceedings of the Eleventh ACM SIGKDD International Conference on Knowledge Discovery in Data Mining*, ser. KDD'05, Chicago, Illinois, USA: Association for Computing Machinery, 2005, 239–248, ISBN: 159593135X. DOI: https://doi.org/10.1145/1081870.1081899. [Online]. Available: https://doi.org/10.1145/1081870.1081899.

20. D. Kelly and N. Belkin, "Reading time, scrolling and interaction: Exploring implicit sources of user preferences for relevant feedback.," Sep. 2001, pp. 408–409. DOI: https://doi.org/10.1145/383952.384045.

21. I. Ruthven and M. Lalmas, "A survey on the use of relevance feedback for information access systems," *Knowl. Eng. Rev.*, vol. 18, no. 2, 95–145, 2003, ISSN: 0269-8889. DOI: https://doi.org/10.1017/S0269888903000638. [Online]. Available: https://doi.org/10.1017/S0269888903000638.

22. R. White, J. Jose, C. Rijsbergen, and I. Ruthven, "A simulated study of implicit feedback models," vol. 2997, Apr. 2004, pp. 311–326, ISBN: 978-3-540-21382-6. DOI: https://doi.org/10.1007/978-3-540-24752-4_23.

23. S. E. Robertson, "On term selection for query expansion," *J. Doc.*, vol. 46, no. 4, 359–364, 1991, ISSN: 0022-0418. DOI: https://doi.org/10.1108/eb026866. [Online]. Available: https://doi.org/10.1108/eb026866.

24. E. N. Efthimiadis, "User choices: A new yardstick for the evaluation of ranking algorithms for interactive query expansion," *Inf. Process. Manage.*, vol. 31, no. 4, 605–620, 1995, ISSN: 0306-4573. DOI: https://doi.org/10.1016/0306-4573(95)00070-W. [Online]. Available: https://doi.org/10.1016/0306-4573(95)00070-W.

25. H. J. Peat and P. Willett, "The limitations of term co-occurrence data for query expansion in document retrieval systems," *J. Am. Soc. Inf. Sci.*, vol. 42, pp. 378–383, 1991. [Online]. Available: https://api.semanticscholar.org/CorpusID:1175919.

26. M. McCandless, E. Hatcher, and O. Gospodnetic, *Lucene in Action, Second Edition: Covers Apache Lucene 3.0*. USA: Manning Publications Co., 2010, ISBN: 1933988177.

27. T. Grainger, *Apache Solr in Action*. Manning Publications, 2014.

28. L. Johnston, J. Kunze, B. Lavoie, and A. Rauber, *Persistent Identifiers: Enabling Services for Data Repositories*. OCLC Research, 2017.

29. N. Paskin, "The digital object identifier: An overview," *Journal of the American Society for Information Science and Technology*, vol. 54, no. 5, pp. 327–329, 2003.

30. D. Robinson, *Content Delivery Networks, Fundamentals, Design, and Evolution*. 111 River Street Hoboken NJ 07030 USA: John Wiley and Sons, Inc., 2017, pp. 183–188, ISBN: 9781119249870.

31. P. Membrey, D. Hows, and E. Plugge, "Content delivery networks," in Jan. 2012, pp. 71–92, ISBN: 978-1-4302-3680-1. DOI: https://doi.org/10.1007/978-1-4302-3681-8_5.

32. Mordor Intelligence. "CDN Market growth, trends and forecast." (Oct. 2020), [Online]. Available: https://www.mordorintelligence.com/industry-reports/content-delivery-market, Download: 24.10.2020.

33. J. Callan, "Distributed information retrieval," in *Advances in information retrieval: Recent research from the Center for Intelligent Information Retrieval*, Springer, 2002, pp. 127–150.

34. W. B. Croft, D. Metzler, and T. Strohman, "Search engines – information retrieval in practice," 2009. [Online]. Available: https://api.semanticscholar.org/CorpusID:2350758.

35. F. Crestani and I. Markov, "Distributed information retrieval and applications," in *Advances in Information Retrieval*, P. Serdyukov, P. Braslavski, S. O. Kuznetsov, et al., Eds., Berlin, Heidelberg: Springer Berlin Heidelberg, 2013, pp. 865–868, ISBN: 978-3-642-36973-5.

36. F. Crestani, M. Lalmas, C. J. Van Rijsbergen, and I. Campbell, ""is this document relevant?... probably": A survey of probabilistic models in information retrieval," *ACM Computing Surveys*, vol. 30, no. 4, pp. 528–552, Dec. 1998, ISSN: 0360-0300, 1557-7341. DOI: https://doi.org/10.1145/299917.299920. [Online]. Available: https://dl.acm.org/doi/10.1145/299917.299920 (visited on 02/15/2023).

37. G. Wiederhold, "Mediators in the architecture of future information systems," *Computer*, vol. 25, no. 3, pp. 38–49, 1992.

38. A. Kopliku, K. Pinel-Sauvagnat, and M. Boughanem, "Aggregated search: A new information retrieval paradigm," *ACM Comput. Surv.*, vol. 46, no. 3, 2014, ISSN: 0360-0300. DOI: https://doi.org/10.1145/2523817. [Online]. Available: https://doi.org/10.1145/2523817.

39. J. Callan and M. Connell, "Query-based sampling of text databases," *ACM Trans. Inf. Syst.*, vol. 19, no. 2, 97–130, 2001, ISSN: 1046-8188. DOI: https://doi.org/10.1145/382979.383040. [Online]. Available: https://doi.org/10.1145/382979.383040.

40. S. Wu, *Data Fusion in Information Retrieval*. 1986, ISBN: https://doi.org/10.1007/978-3-642-28866-1

41. S. Fitz-Gerald, "Information retrieval: Implementing and evaluation search engines, s. buettcher, c.l.a. clarke, g.v. cormack. mit press (2010), £40.95, 606 pp., ISBN: 9780262026512," *International Journal of Information Management*, vol. 32, 496–497, Oct. 2012. DOI: https://doi.org/10.1016/j.ijinfomgt.2012.08.001.

Enterprise Concepts of Information Retrieval

4

The preservation of digital information over long periods of time represents one of the central challenges of the digital age. While analog media such as books or microfilms remain readable for centuries if stored appropriately, digital information is subject to rapid technological change—both in terms of storage formats and the required hardware and software. Therefore, this chapter begins in Subsection 4.1 by addressing the topic of digital long-term archiving, with a particular focus on the role and structure of **metadata**, as these serve as the foundation for the permanent availability, discoverability, and usability of digital resources.

The introduction provides a fundamental overview of the function and significance of metadata in the context of archiving, supplemented by practical definitions and application examples. Following this, in Sect. 4.1.1, various types of metadata such as content, structural, or rights metadata are systematically presented and their respective roles within the archiving process are explained.

A central topic is also the presentation of relevant metadata standards that serve as frameworks for the creation and storage of long-term archival data. These include METS, PREMIS, Dublin Core, LMER, and Z39.87, which are described in Sect. 4.1.2. The various standards address different aspects—from structural description to technical and copyright information.

Special attention is given to the PREMIS data model, which describes the central entities such as objects, events, and agents in a structured manner within the archiving process (see Sect. 4.1).

Section 4.1.3 introduces the Open Archival Information System (OAIS)—a conceptual reference model that serves as the foundation for the design and operation of digital long-term archives worldwide. Here, the structures, information flows, and actors involved in the OAIS model are explained, including the central concepts such as SIP, AIP, and DIP.

M. Hemmje and S. Wagenpfeil, *Multimedia Information Retrieval*,
https://doi.org/10.1007/978-3-662-73310-3_4

Finally, Subsection 4.2 bridges to the topic of Semantic Information Retrieval, highlighting the integration of semantic technologies into modern archival systems. Semantic enrichment and querying of archived content open up new possibilities for more intelligent knowledge discovery and linking.

This chapter thus provides a comprehensive overview of the theoretical and practical foundations of digital long-term archiving and enables readers to critically classify key terms, models, and standards and to apply them to their own scenarios.

4.1   Long-Term Archiving

In the context of long-term archiving (English: "Preservation"), metadata play a particularly important role. Therefore, this subsection first introduces a series of considerations based on the relevant literature before moving on to present this field using practical examples.

The PADI initiative (Preserving Access to Digital Information) of the National Library of Australia [3] refers to metadata for long-term archiving as "structured methods to describe and record information needed to manage the preservation of digital resources." Furthermore, it defines preservation metadata as "a subset of administrative metadata that supports the management of information, as well as technical metadata that enables current access to digital content." Dr. Rohde Enslin from the Institute for Museum Research in Berlin describes metadata for long-term archiving as "a continuously updated record of the history of changes to a file. Requires monitoring and recording of all changes" [4]. Priscilla Caplan defines metadata for long-term archiving as all "data necessary to be able to retrieve a document at some later time" [5].

Metadata for long-term archiving thus accompany the entire archiving process of digital materials and are necessary to ensure the long-term preservation and renderability of digitally archived material. The reason and necessity for this lie in the fact that multimedia assets are generally highly dependent on technology, and these technologies are subject to continuous change. In other words, as technological development in hardware and software continues, it must be ensured that files and documents can still be found, accessed, read by a computer system, and displayed to a user using new technology.

Just as rapidly as storage media change (e.g., datasette > 5 1/4′′-inch diskette > 3 1/2′′-inch diskette > CD > DVD > USB stick > network storage > cloud), so too do the file formats in which files are stored on such media change. These formats are subject to ongoing development, as the tools for creating documents and other digital information objects are also constantly evolving. Manufacturers generally strive for a certain degree of backward compatibility in their document formats (as with Adobe PDF) or at least in the programs used to open and edit these documents (see, for example, Microsoft Office 2007 with the new Office Open XML standard, which can also open old Office formats). This area and its impact on metadata for long-term archiving must therefore be given special consideration.

In this chapter, we therefore aim, in the context of digital long-term archiving, to introduce the topic of metadata and, in particular, to highlight the role of metadata for long-term archiving in supporting the archiving process. To address the aspects mentioned, various standards for metadata for long-term archiving have also been established. Therefore, this chapter will also provide an overview of some of these standards. In particular, standards have been selected that either relate to specific file types and formats, such as the Z39.87 format [6], or those that—like the PREMIS data model [7]—are quite general in nature. The PREMIS Data Dictionary will therefore also be described in more detail [7].

To understand what is meant by metadata for long-term archiving, it is first necessary to define and explain the concept of metadata in general. In the words of Tim Berners-Lee, the founder of the World Wide Web, "Metadata are machine-readable information about web resources or other things" [8]. Web resources are generally documents such as HTML pages, i.e., digital data and other machine-readable information objects. These are, without question, digital data sets. Thus, metadata can also be described as data that describe other data, leading to a widely accepted and common definition: "Metadata are data about data." This can be quickly illustrated using the example of digital photography (see also Subsection 2.2). The actual data generated when taking photos are, of course, the media data of the photos themselves. In digital cameras, these photos are stored digitally as pixel data in the form of color values (e.g., in JPG or TIFF format). In addition to these data, today's cameras usually automatically capture a variety of additional information. These additional data are often stored in the Exchangeable Image File Format (EXIF). Some examples of metadata within EXIF data include date and time, technical image format information (image size, resolution, color depth, etc.), camera used, and camera settings (exposure time, ISO value, etc.). This is also a good example to highlight two further aspects of metadata. First, these are automatically captured metadata, which in this case are recorded and stored directly and fully automatically by the camera during data creation. Modern object recognition methods can also extract further multimedia features from multimedia assets and thus provide higher-level, semantic information as metadata (see Subsection 2.8). These metadata can be embedded in the original file as additional information alongside the actual data (e.g., EXIF). Other examples of embedded metadata include similar additional information in Microsoft Office documents (Word, Excel, PowerPoint, PDF) or in music files (e.g., in MP3 format).

Data carriers from the past 50 years and the information stored on them are undoubtedly more endangered than ever before since the invention of writing by the Sumerians in antiquity. Fifteen years can have a more devastating effect on digital media than over 3000 years on clay tablets and papyri or 500 years on printed media.

The techniques for extracting and interpreting data are being replaced by new methods at ever shorter intervals. They are tied to technological development and its cycles. Opening a book, for example, is easier than reading a compact disc. Such a rapid pace of change in cultural technologies for encoding knowledge is unprecedented in human history. The situation is further exacerbated by the frequent existence of a multitude of competing methods (such as the recent Blu-ray and HD-DVD), which further complicates matters. Critics therefore sometimes speak of a digital Middle Ages that we risk entering if we do not effectively protect our knowledge against the loss and obsolescence of encoding and storage methods. Considering the scale at which information is currently produced and stored exclusively in digital form, the advantage of proven methods, well-founded decisions based on objective criteria, and careful planning becomes obvious. A single mistake can lead to the irretrievable loss of cultural assets and knowledge. The task of preservation planning addresses this field.

Metadata play a crucial role both in preservation planning and in the storage of multimedia assets. There are automatically captured and manually captured metadata. These can be embedded in the multimedia assets, if the respective file format allows it, or stored externally (e.g., in databases, associated metadata files, etc.). Manual capture or supplementation of metadata is always necessary when either automatic capture has not worked or was incomplete. For example, automatic content description of videos is still associated with major challenges today and only works in a narrowly defined thematic context. Here, the capture of metadata by archivists is still common practice. For the capture of subjective metadata (such as evaluation and mood of image content), no automation is currently possible.

External metadata refers to information that is stored separately, usually in other files or databases. An example of this is the Metadata Encoding and Transmission Standard (METS) [9], which will be described in more detail later in this chapter. Document management systems, which, in addition to pure data storage, generally also have a database in which further information can be stored for each document, also make significant use of external metadata. The following section will first describe the relationship between metadata and digital long-term archiving, as well as present some further relevant possibilities for classifying metadata.

4.1.1 Types of Metadata

To illustrate the role of metadata in digital long-term archiving, the various types of metadata for long-term archiving will now be presented in more detail. There are many different attempts to classify metadata by type, but it is difficult and ultimately not meaningful to draw exact boundaries. For the following overview, a classification has therefore been chosen that highlights the most important types of metadata in the context of digital long-term archiving [10, 11].

Content metadata, often also called descriptive metadata, are similar to the classic catalog or bibliographic data in libraries or museums. They contain information about the title, author, publication date, as well as keywords and the subject area to which the publication belongs. However, bibliographic metadata often also contain technical information, such as the format and material of a work. For digital content, however, these are usually referred to as technical metadata and are thus considered separately.

Technical metadata thus also implicitly describe the required infrastructure (hardware and software) necessary for the faithful opening of a document. This includes fonts, color profiles, file format, and other media-dependent types of information.

Structural metadata contain information about the structural composition of a digital object. The simplest way to illustrate this is with a website (consisting of several HTML files and other resources), where, in addition to the individual files, their storage location (folder structure) must also be saved. It may also be necessary to logically link publications consisting of several separately stored documents via a kind of table of contents. However, the original structure must be preserved and maintained for this purpose.

Administrative metadata provide information about processes carried out in the course of managing content. This also includes changes to a file, with details about the original version and information about the authenticity of the data.

Rights metadata, or copyright and usage rights metadata (English: "Intellectual Property Rights," IPR), contain information about the use of archived data. Digital Rights Management (DRM) plays an important role here when it comes to who is allowed to see what, when, and how often.

> Metadata play a crucial role in long-term archiving.

4.1.2 Essential Concepts, Properties, and Relevant Standards

All these types of metadata for long-term archiving can be recorded and stored in any structured form. Before starting to collect and maintain collections of metadata documents, it is necessary to consider which conventions regarding the structure of the metadata and their indexing should be followed. Specifically, this means thinking about which metadata should be collected at all, and how and where they should be stored.

Various standards have now been established to address these questions. While they cannot completely relieve users or developers of these considerations, they do provide a certain framework within which to operate. In particular, if a later transfer of the objects with metadata to another archiving institution is envisaged, or if consolidation with such an institution may become necessary, compliance with standards for metadata in long-term archiving is indispensable to avoid manual post-processing. Standards generally ensure that the important metadata and the structure in which they are represented are present. Thus, standards serve completeness and interoperability. Standards also provide guidelines or suggestions on how this metadata should be stored. Some of these standards are presented below.

The **Metadata Encoding and Transmission Standard (METS)** [9] specifies that metadata should be stored in an XML-based format. METS was developed to manage digital objects within a library as well as to ensure the exchange of these objects with other libraries or users. As a generic container format, METS can in principle be used for all data types and formats. The seven main sections of a METS XML file include, among others, descriptive (i.e., content-related), administrative, and structural metadata. Technical and rights metadata are also included here as part of the administrative metadata. Thus, the types of metadata mentioned above are fundamentally covered by METS.

With the **Dublin Core Metadata Element Set** [12], a standard for the description of resources (e.g., HTML pages) using metadata is gradually being established on the Internet as well. The Dublin Core Metadata Element Set consists of at least 15 elements, i.e., the Simple Dublin Core, which is recommended by the Dublin Core Metadata Initiative (DCMI). These elements can also be categorized according to the types mentioned above as follows: content metadata (e.g., title, subject, description), technical metadata (type, format), structural metadata (source, relation), administrative metadata (date), and rights metadata, with a primary focus on content metadata and only partial coverage of the other types. Dublin Core metadata can be embedded in the document (as with HTML pages or other XML-based documents) or stored in external databases. This is necessary for multimedia objects such as images, but even for text-based documents, centralized storage of metadata can be advantageous.

As a representative of the group of standards for technical metadata, the **Standard for Long-Term Archiving Metadata for Electronic Resources (LMER)** should also be mentioned [13], which, like most modern metadata standards, is based on an XML schema. LMER was primarily conceived as an exchange format; storage was not considered. Thus, storage can take place in any XML structures of other standards. LMER does not claim to be comprehensive. In particular, metadata that depend on a specific file format are not considered. Instead, the XML schema provides for the addition of any further XML structures for format-dependent descriptions. This modularization is a major advantage of this standard, as it allows future requirements of new file formats to be met through appropriate further development of the modules.

The German National Library, the Göttingen State and University Library, the Gesellschaft für wissenschaftliche Datenverarbeitung mbH Göttingen (GWDG), and IBM Deutschland GmbH have developed a solution for the long-term preservation and availability of digital documents as part of the kopal project. The result is the **Universal Object Format (UOF)** [14], which is based on METS and LMER and is specifically tailored to IBM's proprietary DIAS archiving system. Each document in UOF is described using a METS-compliant XML file (mets. xml), which must therefore be valid according to the METS document type definition. This XML file provides the basic framework for all required metadata. The administrative and technical metadata are covered by LMER. For descriptive metadata (content metadata), any XML structures can be included (including Dublin Core).

The standard for **Machine-Readable Cataloging (MARC)** is a very comprehensive metadata format (including for content metadata) [15], which originated in the 1970s from an initiative of the Library of Congress. Many formats based on it now exist, harmonized by the MARC 21 standard. The **Metadata Object Description Schema (MODS)** is also a simplified and shortened form of the content metadata from MARC 21 [16]. In terms of the amount of metadata, it is positioned between Dublin Core and MARC 21. The **Metadata Authority Description Schema (MADS)** is also based on MARC 21 [17], but only describes the authority part, i.e., people, institutions, events, and conditions.

If one combines the types of metadata mentioned so far, one already gains some insight into the classification of metadata for the archiving and preservation process, in short, preservation metadata, within the overall construct of metadata, especially in connection with technical and administrative metadata. In fact, all technical metadata are also preservation metadata, and they make up a large part of it. However, these two types are not completely congruent. Preservation metadata also store information about the archiving and preservation process itself. This includes, in particular, necessary changes to the data caused by migrations or other preservation actions on the digital objects. These data represent specific information about the preservation and archiving process, which are shaped by administrative metadata. Furthermore, the collection and maintenance of preservation metadata, in contrast to many other types of metadata such as content-related information, is not a one-time action. When changes are made to the data to be archived or to the technical environment, these must be reviewed and revised to ensure that the data remain securely retrievable and displayable in the future. All changes made must be documented and recorded as a history. Further important aspects of preservation metadata are described by way of example in the next subsection. The work by Priscilla Caplan, created as part of the DCC Curation Manual, is referenced here [5], in which some examples of commonly used elements of preservation metadata are named and described. In addition, reference is made to the PREMIS data model, which will also be presented later in this chapter.

As the most important aspect of preservation metadata, the format properties of a file, such as the MIME type, should first be mentioned. Unfortunately, this is not always precise enough to accurately reflect the format of a file. For example, consider the variety of audio and video codecs that can be used within a single MIME type. In addition, preservation metadata always include the significant properties of an object. These indicate which important properties of a document (e.g., only the text portion) must be preserved during migration or emulation to ensure the document remains usable. Furthermore, preservation metadata are collected about the usage environment. These are essential for the later display of archived materials using the associated tools and their user interfaces. This includes data about hardware, software, and supplementary files required to display the archived data. There is no point in archiving database contents without knowing how the individual tables are linked within the overall schema and how they interact in different display views. The same applies to the general availability of a file if it is unknown which program can open the file and on which types of computers this program runs.

Furthermore, with regard to the persistence of an object, it is important for preservation metadata to know whether the archived document corresponds to the originally archived version or whether it has been changed in the meantime. In practice, this is often done by generating a digital checksum and comparing it to the original document and to each modified version, with the checksum then stored together with the document. Within preservation metadata, the technical metadata already mentioned are also of great importance. Many technical metadata are indeed independent of the file format. Thus, information about size, format, and persistence can always be collected and stored. The PREMIS data model deals with these general technical metadata. Other information only arises with certain file formats and therefore cannot always be collected. These are not considered in PREMIS. There are standardization efforts for many formats. One of the most thoroughly developed standards is the NISO standard Z39.87, which describes the technical metadata for digital images. In addition, for each document, the provenance, any changes to the document itself, and important events (e.g., format migrations) are usually stored within preservation metadata. Sometimes an entire change history for an object is recorded. In PREMIS, however, archived objects cannot be changed. Any change results in a new document, which is linked to the original version. In all cases, however, the actions performed on the document are recorded. In order to link the metadata to the object to be archived, this information must be collected and attached to the document as a complex annotation object in a structured manner. The METS data model has established itself as the de facto standard for structuring and transmitting preservation metadata.

The motivation for having standards in the context of preservation metadata is again found in the goal of a universally consistent approach to archiving. Only in this way is it possible to merge multiple data holdings from decentralized archives and to benefit from the experiences of other users of the same standard, for example regarding migration and preservation of data. Among existing metadata standards, there are those that are independent of the type of object to be archived, and others that are only applicable to a specific format, since the characteristics of individual formats can differ greatly. For example, it is neither possible nor meaningful to determine the dimensions or color depth of a plain text file. It is, of course, also legitimate to ask why there are so many different standards. This is, however, relatively easy to understand when one recalls the many different possible media contents and their diverse file formats as well as their fundamental properties and peculiarities. What use is a comprehensive standard for describing image data if only plain text documents are to be archived?

As a representative of format-specific standards, the NISO standard Z39.87 has already been mentioned, which includes a very comprehensive data dictionary for describing digital images and image collections in a system-independent manner. Originally developed for TIFF formats, this standard is, due to its flexibility, also very well suited for JPEG-2000 and DNG ("Digital Negative").

This standard covers technical information (metadata) relevant to the management of digital still images. This includes tasks related to quality assurance, data processing, and long-term preservation of images. Rights metadata, for example, are not covered, but are addressed by PREMIS. Some (general) elements overlap in PREMIS and Z39.87, so efforts have been made to harmonize these elements to avoid conflicts or duplicate recording. The Z39.87 data dictionary was adopted by the Library of Congress and further developed in the NISO MIX standard [2] so that an XML schema was developed to store the required metadata in XML-compliant form. As one of the most important representatives of a standard for generic preservation metadata, the following section will discuss some general aspects of PREMIS. The abbreviation PREMIS stands for the **Preservation Metadata: Implementation Strategies Working Group** [7], a working group founded in 2003 by the Online Computer Library Center (OCLC) [18] and the Research Libraries Group (RLG) [19], which has since merged with OCLC. The aim of the merger was to continue the work jointly within the Preservation Metadata Framework Working Group (2001 to 2002). This OCLC/RLG working group essentially dealt with fundamental issues regarding metadata to support digital long-term archiving, taking into account the OAIS reference model [20]. Members of this working group included staff from institutions and projects that had already made a name for themselves in this field. Examples include NEDLIB (Networked European Deposit Library) [21], CEDARS (CURL Exemplars in Digital Archives), the National Library of Australia, and Harvard University [22]. After the results of the OCLC/RLG working group were available in the form of a framework in 2002, the PREMIS working group was established, which was initially tasked with conducting a large survey to determine the current state and needs of libraries, archives, and companies involved in digital long-term archiving. Based on the results of this survey, a data dictionary was designed and developed, containing the most important preservation metadata. This data dictionary was completed in May 2005. Thus, the work of the PREMIS working group was also completed. However, the work was not finally concluded, but has since been continued by the Library of Congress, which has taken over both the further development of the data dictionary and a number of other topics related to PREMIS.

The PREMIS data model was designed to describe the logical relationships between the metadata elements used (so-called entities). Figure 4.1 illustrates these relationships. The direction of the arrows here indicates a possible linkage.

Objects in the PREMIS data model are fundamentally the elements to be archived, with a unique identifier, information about the file (size and format), information about the required hardware and software, and other relevant properties. Relationships to intellectual entities, events, and rights are also recorded. The PREMIS data model also describes three basic types of objects: file objects, bitstream objects, and representation objects. File objects are known to an operating system and are defined by their name, file format, access permissions, and file system information such as size and modification date. A bitstream object is part of a file that is significant for archiving purposes. Bitstreams cannot easily be converted into a file without adding additional information (e.g., header data). An example of this is TIFF files consisting of multiple images, or binary files serialized in XML files, as well as embedded metadata. Representation objects group several files into an intellectual entity. For this, both the participating files and the structural dependencies are recorded. Examples include websites with multiple images, where the placement and size of the images are important, or articles or books consisting of several files, each containing a single chapter or where image and text are separated.

An **Intellectual Entity** is thus a bundle of related content or objects that describes a unit. This can be a book, a map, a photo, or a database. An intellectual entity can also contain another intellectual entity, for example, when a website displays an image, which in turn is an object. There are one or more digital representations for an intellectual entity. For example, a photo can be stored both as JPG and as TIFF. Intellectual entities are described by content metadata, which are not part of PREMIS. Other metadata standards can be used here, such as METS or Dublin Core. The events are within PREMIS actions, which are related to an object or agent and concern the archiving process. This refers, for example, to information about changes to a file (change history) or simple validity and integrity checks on objects, as well as the results of these changes or checks.

Routine backups can also be recorded here, but it is up to the archiving institution to decide whether this level of detail is required. The **Agent** in PREMIS is a uniquely named, event-triggering, and identified person, organization, or software. The role of an agent depends on the associated event and is therefore also recorded in the **Event Entity**. The **Rights** in PREMIS define which agents are allowed to perform certain events on associated objects. These rights can also be prescribed by law and relate to copying, conversion, or migration.

The PREMIS data model also describes two basic types of relationships: relationships between objects and relationships between different entity types. Object relationships can be defined by structure or by derivation. An example of a structural relationship is the "part-of" relationship (e.g., an image is part of a website). A derived relationship arises, for example, from a migration event of an object ("child-of" relationship): the migrated file is the child of the original file.

Entity type relationships were already shown in Fig. 4.1 with the arrows and were introduced during the description of the entity types. For example, an object

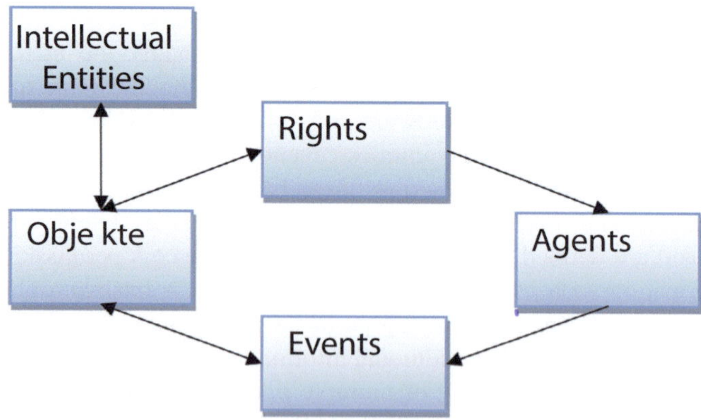

Fig. 4.1 The PREMIS data model

has relationships to intellectual entities, rights, and events. An agent has a connection to an event.

The PREMIS data model finds its formal representation in the PREMIS Data Dictionary. The PREMIS Data Dictionary contains semantic units for objects, events, agents, and rights. Since intellectual entities are already covered by their content metadata, they are not considered here. Semantic units describe properties of an entity, i.e., important information essential for planning and carrying out preservation. The current PREMIS version dates from 2015 and is being further developed, optimized, and adapted to current developments as an international standard for long-term archiving.

> There are various standards for long-term archiving.

4.1.3 The Reference Model for Open Archival Information Systems

Nowadays, very large amounts of digital information exist in various forms. Parts of this information represent the knowledge of individuals, institutions, or systems. Using this knowledge, decisions can be made, new information can be derived within knowledge systems, and so on. In these application contexts, the terms knowledge production and utilization processes, knowledge collections, and knowledge life cycles are now commonly used. The knowledge that already exists should continue to be available for as long as possible in the future, and should also remain accessible and usable. Therefore, knowledge is of great importance to all societies. Explicit knowledge exists in various forms. In written form, it is found in books, newspapers, and publications. In audio form, it is found in music, audiobooks, radio broadcasts, etc. In visual form, it is also found in images, films, reports, documentaries, as well as in drawings, sketches, diagrams, and so on. This list could be extended indefinitely. Accordingly, the challenges in long-term archiving are at least as diverse as the fields of multimedia and multimedia information systems themselves. Many of these information objects already exist in digital form or are gradually being digitized (such as large collections of national libraries). Digital long-term archiving deals, among other things, with the question of how such digital content can be securely stored for long periods, effectively retrieved, accessed, and efficiently used.

If we now consider the media or formats used to store this information, it quickly becomes apparent that different approaches are chosen. For some, it is sufficient to store the digital content on storage media in its original format, convert text files to PDF or XML format, or set up a database for various data types. As long as individual institutions or people do not wish to access, integrate, or further process the knowledge of other systems automatically, such a type of archiving initially poses no problems. However, it is easy to find more advanced scenarios in which the information should be available in a more uniform form for external access. This is done, for example, in some online catalogs of libraries. In the field of space exploration, large amounts of digital data were already being generated in the 1960s. This created a need to archive this information uniformly, securely, and for the long term.

The Consultative Committee for Space Data Systems (CCSDS) [23] therefore began developing a draft standard in 1997 for a unified conceptual architecture for archival information systems within a reference model called the **Open Archival Information System (OAIS).** It was not until 1999 that the conceptual OAIS system model developed up to that point was published as a Red Book. In 1999, this Red Book was also submitted for standardization to the International Organization for Standardization (ISO) for the first time. This draft was recognized by ISO in 2001 as the international standard ISO 14721. The aim of this standard is to create a common understanding of an architecture. In the following, this section presents the conceptual OAIS model of an archival information system that underlies the standard.

Since then, many documents in organizations have been digitally archived and collected using OAIS-compliant systems. Through a standardized access method, different documents and any digital data (video, audio, image material, etc.) can be made uniformly available to the public for free access. For accessing data published using a unified architecture, the **Open Archives Initiative (OAI)** developed the Protocol for Metadata Harvesting (OAI-PMH) [24]. However, OAIS should not be confused with OAI-PMH. The purpose of OAI-PMH is to publish documents uniformly for different user groups. The main use of OAI-PMH is at universities for the publication of scientific documents. The following sections will now present the components of the OAIS model in more detail.

OAIS thus defines a conceptual framework model for a system for archiving electronic documents. This first describes the basic terminology as well as the main tasks of such a system and the guidelines for its use and handling. The aim is to create a common basic understanding of the subject for global information and experience exchange. The focus is on defining a general, conceptual, theoretical, or semantic model for the functionality of an archive. The architecture of the OAIS model consists of three associated models: the OAIS process model, the OAIS information model, and the OAIS data model. Within OAIS, long-term archiving is first considered from the perspectives of the information and process models. The information model is described as follows: In the information model, the actual content data is separated from the descriptive information, the metadata, so that they can subsequently be stored separately. The data here refers to, for example, texts, program code, video material, etc. The semantic content of the data in OAIS is achieved through an additional architectural component, the Knowledge Base. This is needed to be able to understand the data. To be able to fully interpret the data in an archive, OAIS defines a third component, the Representation Information.

An example of this division can be seen in the source code of a program. The data in this case is the given source code. The Knowledge Base contains knowledge about the programming language used, and the Representation Information encompasses further knowledge about this programming language, which could be available, for example, in the form of a textbook, script, or other description of the syntax and format of the programming language. Thus, the information model of the OAIS reference model consists of the following components and relationships (see also Figs. 4.2 and 4.3):

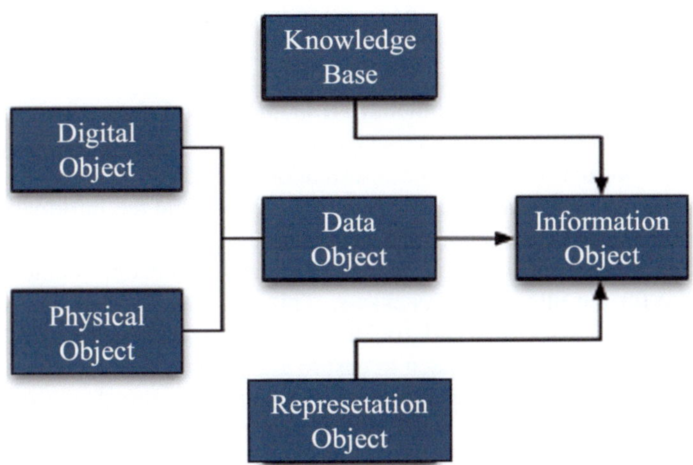

Fig. 4.2 Information model of the OAIS reference model [23]

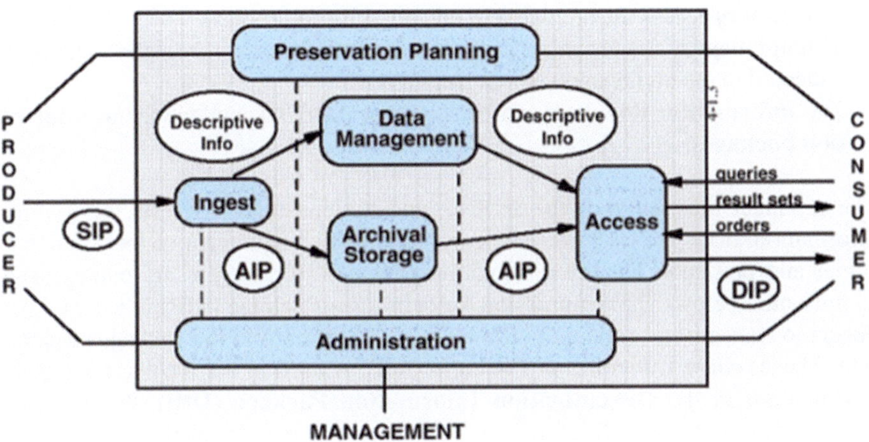

Fig. 4.3 The components in the OAIS reference model [23]

- Digital objects and physical objects are the **data objects** (English: "Data Objects"). Representation information is required for the evaluation and processing of this data.
- The **Knowledge Base**, the data objects, and the representation information together form the **information objects** (English: "Information Objects").
- So-called **packaged information units** (English: "Information Packages") are units into which the information can be split and which are used for transmission. Information Packages can be divided into content (English: "Content Information") and preservation description information (English: "Preservation Description Information").
- **Content Information** includes the data and the representation information.
- **Preservation Description Information** includes the information required for archiving and integrating the content information. This includes, among other things, the original source of the information, the relationship of the information to other information objects within the archive, and the information necessary to ensure the preservation of the integrity of the information.
- **Descriptive Information** includes the metadata (so far only content description data) for the information packages. This metadata is needed to search for the information packages and to interpret them. This includes information about the creator, measures taken to ensure availability, the time of creation, the unique identification of the document, and the requirements that must be met to display the document.

To ensure the integrity of the archive in a global environment, OAIS uses a three-layer component structure with the following components (see Fig. 4.3).

- **Producers** create the content and store it in the archive.
- **Consumers** access the archive to read and reuse the content.
- **Management** is the organization that determines which information should be collected in the archive and which should not.
- The **information flow** between these components takes place using **information packages**.

The producer thus supplies the archive, and the consumer requests the required information from the archive. There is no direct communication between consumer and producer. For the information packages, three types are distinguished as presented below. The **Submission Information Package (SIP)** is used by producers to transmit information to the archive. The format of the packages is specified. The **Archive Information Package (AIP)** is used in the archive to store the information. In the **Dissemination Information Package (DIP)**, the consumer receives the requested information in packages. The format for these packages is also specified. Within the archive, a transformation takes place between these three package types. The processes required for this are described in the OAIS process model. The conversion of packages from one type to another is a complex task and is carried out in several steps. The necessary processes run within an OAIS and are responsible for generating the dissemination information packages from the submission information packages. These are shown in Fig. 4.3 and are explained in more detail below.

Common services include operating system-specific services, communication between multiple processors, network services, name services, management of temporary storage, error handling services, and services to ensure security.

The **ingest process** receives the SIPs from the producer. Its tasks include storing this information in the system and preparing this data for further processing in the internal format. It is easy to imagine that converting data from different formats into a uniform format can often prove difficult and complex. To reduce the effort, certain formats for the delivered data are usually agreed upon with the producers. This greatly minimizes the number of possible formats and schemas that need to be supported. These agreements are also usually managed by the ingest process, and incoming data is checked for compliance. Furthermore, the data is validated by the ingest process for completeness and correctness. To ensure authenticated access to the system, a producer must authenticate themselves to the system, thereby creating a control mechanism for access. Performing authentication is another task of the ingest process. The verified SIP is then converted into an AIP, and the associated descriptive information is generated from the SIP. The generated descriptive information is then stored in data management.

The **archival storage process** is responsible for managing, storing, and providing the data in its original state. Archival storage has various storage media available for storing this data. This data is archived for long periods and must therefore be managed efficiently. This requires long-term and well-conceived storage management. Archival storage must also guarantee the integrity of the data throughout the entire period. For this purpose, the storage medium is periodically refreshed, as its lifespan is limited. Archival storage is responsible for supplying access with the requested data.

The **data management process** maintains the descriptive information generated by the ingest process from the SIP. Unlike archival storage, a database is always used for storing the data. Managing this database is the task of data management. Furthermore, the delivered data must be prepared for storage. Queries for information can only be made via data management, as only it has access to the descriptive information for all the data.

Administration is responsible for the functionality of the archive. It negotiates criteria with the producers that must be met. These criteria are also checked by administration. Furthermore, this process constantly monitors the operations in the system. The basic components of the archive are the hardware and software. These must be maintained, renewed, and constantly monitored. This task is also performed by the administration process. To achieve fast response times and good storage utilization, the entire system is regularly optimized.

The task of the **preservation planning process** is the long-term oriented archiving of the data, so that access to this data is still possible in the future. Thus, the tasks of this process fall within the field of long-term archiving. For this purpose, current changes and trends in hardware and software on the market must be monitored in order to derive and implement necessary actions in a timely manner. One such action can be data migration. This may result in the need to migrate data to a new system (using servers from a different manufacturer) or for the system to support new formats (e.g., data should be deliverable in standardized XML formats).

Access is the process that forms the link between the system and the consumers. It provides the interface to the users, receives search queries, and prepares the results obtained from archival storage according to the interface description. Furthermore, this process monitors the correct delivery of generated information to the consumers.

These are the six processes that represent the information flow within an archive. These processes are formally defined in the reference model, as experience has shown them to be necessary. OAIS does not define concrete proposals or guidelines for implementing these processes, as OAIS is a reference model and not a reference implementation.

In [10], further properties and possible requirements for an OAIS are listed. Possible implementation alternatives are given for the resulting requirements. For example, the interoperability of archives is considered from the perspective of a user or a producer. A user is interested, for example, in cross-archive searches and in uniform interfaces. For instance, it is possible to search the data holdings of affiliated libraries via a library. A producer is also interested in a uniform interface and in the simultaneous publication of data in several archives. From these requirements, four proposals for organizing archives result:

- In an independent archive, the data holdings are managed locally. Users can only access the data within this archive. The interfaces for publication and for obtaining data apply only to this archive.
- In cooperating archives, data is exchanged between them. For this, an archive must know the DIP specification of the other archive. A user can thus find data from another archive when searching in one archive. A producer can thus make data accessible in several archives with a single publication.
- In federated archives, a user can search multiple archives simultaneously. Data is exchanged between the search platform and the individual archives. A producer also only needs to publish their data in one archive.
- Several archives can share a data system for data management. In this case, the data base would be the common component.

Some of these considerations can be found in distributed database systems. OAIS also addresses the necessity of data migration. This may be required due to technological advances in hardware or software. If data is migrated to other systems, the AIPs must be updated. The updating and conversion of AIPs to a new data structure can be caused by new technical requirements. Whether it is the migration of the actual data, the migration of metadata, the adaptation of AIPs, or the updating of hardware or software, the integrity of the data and the entire archive must be maintained.

Currently, various OAIS-compliant products and solutions are emerging or already available on the market. These range from commercial solutions to open-source projects. There are also projects that support the introduction of an Open Archival Information System. These products and projects include, among others, Kopal [25], DigiTool [26], DIAS [27], Fedora [28], EPrints [29], MyCoRe [30], DSpace [31], and others.

Kopal is a system that complies with the ISO-14721 standard. This system was developed based on DIAS-Core (IBM). DIAS was extended by the Göttingen University Library and the German National Library with additional functionalities. These extensions make it possible to capture and use metadata and archival data for searching. In conformity with the OAIS reference model, an access component was defined to provide users with data. For this requirement, the Access Tool was implemented in Kopal, enabling structured access to the data holdings. The Ingest Tools are used for incorporating data into Kopal (these are also specified as necessary by OAIS). These tools support the compilation and preparation of data from which the AIP is generated. The format for SIP contains the following metadata: NBN, originalNBN, refPlatformNBN, supplierName, starterFileName, dateOfCreation, ingestUserID, sourceType, setupFileName, sourceDescription. The specification of the SIP is described in kopalDIASSIPInterfaceSpecification.pdf. There is also a specification for DIP.

The Bavarian State Library, together with the Leibniz Supercomputing Centre, developed the BABS system, which can be used for managing and archiving digital documents of any kind. The databases, management systems, and archival systems used in BABS are presented in [1]. Part of the overall system is the DigiTool tool, which is developed by the company Ex-Libris. DigiTool is mainly used at European and American universities for managing, searching, and publishing digital content. Further descriptions of ExLibris and DigiTool are available on their homepage.

Fedora (The Flexible Extensible Digital Object and Repository Architecture) is an open-source project of Cornell University, Department of Information Science, and the Library of the University of Virginia. Fedora is a library for digital content. Access to a Fedora system can be achieved using various technologies (web services, web platform). The metadata description of the data is done in XML using Digital Object Relationships. These make it possible, among other things, to manage, group, and create references between data. This enables the establishment of semantic relationships between data, on which various advanced access methods can be built.

In several OAIS implementations (such as the Kopal version of the Koninklijke Nationalbibliotheek of the Netherlands called E-Depot), DIAS-Core (Digital Information Archiving System) forms the core of the overall system. DIAS manages the stored data and provides an environment for different preservation strategies. Multi-tenancy, precise import and export interfaces enable the use of DIAS in different environments. The system can be integrated into various workflows and institutions. By using proven standard software components (such as IBM DB2 Content Manager, IBM WebSphere Application Server, and IBM Tivoli Storage

Manager), long-term stability, performance, and scalability are ensured. Metadata is stored in the Library Server of the DB2 Content Manager, and the objects are stored in the Resource Managers of the DB2 Content Manager. IBM DB2 Content Manager has an interface to IBM Tivoli Storage Manager, which supports storing data on different media.

The OAIS reference model thus describes a complex conceptual system architecture that can be used in the design and implementation of an archival system. Whether all the modules listed in OAIS must be implemented as separate modules in an archival system depends on the scope of the respective archival system. Due to the abstract definition of the OAIS reference model, it can be implemented in various industries. Since all the necessary components are fully listed, there is little risk that important subcomponents could be forgotten when designing one's own archival system. In this respect, there is no need to reinvent the wheel. This also applies to the OAIS reference model, whose design involved various institutions with great expertise (both practical and theoretical) up to 2003. The fact that it even became an ISO standard ISO-14721:2003 shows that this concept is internationally recognized. However, not every company wishing to develop an archival system can afford the effort required for the analysis and design of such a model. It is better and more cost-effective to benefit from and learn from the official and published experiences of others. Of course, it may happen that when introducing OAIS in an existing institution, the current circumstances and conditions are not compatible with the model. In the long run, however, it is advisable to adapt one's own processes and switch to a standard. Overall, technical specifications, reference implementations, and certification test software for OAIS-compliant developments would certainly also be useful and beneficial, as they would support and facilitate the emergence of OAIS-compliant products and technologies.

> OAIS is a reference model for long-term archival systems.

4.2   Semantic Information Retrieval

In Subsection 2.8 the basics of semantics were introduced. In this subsection, we will now look at how this can be concretely used for information retrieval based on the models and concepts from the previous subsections. This usage refers both to the semantic capture of document content and to the semantic interpretation or annotation of arbitrary multimedia features.

Enriching queries with external data sources—for example, in the sense of taxonomies—is useful to expand the elements of a query with related terms (e.g., by considering homonyms). However, semantic search goes far beyond simply expanding a query with additional terms. The following subsections are based

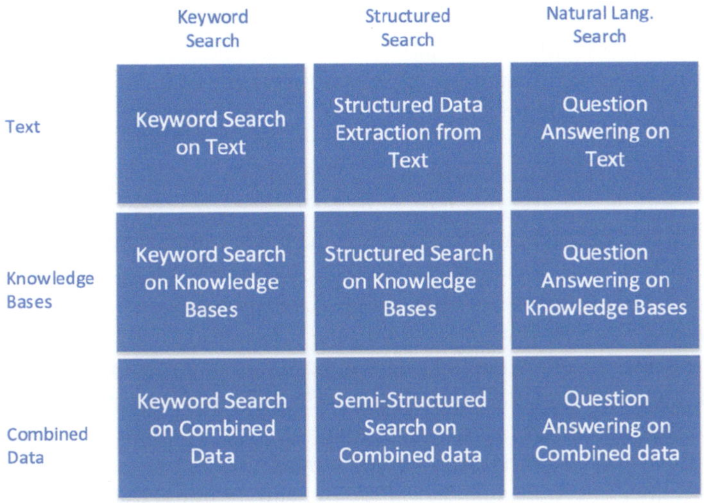

	Keyword Search	Structured Search	Natural Lang. Search
Text	Keyword Search on Text	Structured Data Extraction from Text	Question Answering on Text
Knowledge Bases	Keyword Search on Knowledge Bases	Structured Search on Knowledge Bases	Question Answering on Knowledge Bases
Combined Data	Keyword Search on Combined Data	Semi-Structured Search on Combined data	Question Answering on Combined data

Fig. 4.4 Levels of semantic search

on Bast [32], which provides a very structured overview of the various facets of semantic search. They distinguish approaches to semantic search on two interwoven levels. On the one hand, there is the level that indicates how a query is passed to semantic search (the search paradigm). On the other hand, there is the level of the data on which the search is performed (see Fig. 4.4).

At the data level, the following three types are distinguished:

- **Text:** Refers to a collection of documents with textual content or the textual representation of multimedia features—as introduced in Subsection 1.6.4.
- **Knowledge Bases:** Here, this essentially refers to the provision of information via ontologies, as documented in Subsection 2.8.1. The ontologies are represented using Semantic Web languages and made available for access via triplestores.
- **Combined Data:** Both text and knowledge bases are ways to express knowledge. Text is probably the most intuitive form of expression for humans (due to its everyday use). However, text may not be written unambiguously. Knowledge bases, on the other hand, are particularly well suited for making precise statements due to their formal nature. A hybrid of both approaches can be an ideal complement for the most precise and understandable representation of a subject. There are basically two approaches to implementing such a hybrid representation. One is the introduction of links in a text that refer to specific resources in a knowledge base. The other is a type of linking that generally refers to a set of different knowledge bases. In the latter case, it is of course also possible for resources to be described using several (different) ontologies.

The following three types are distinguished as search paradigms:

- **Keyword Search:** Searching by specifying keywords is probably the best-known way to communicate an information need to an IRS. All best-match IR models implement this. The widespread use of this paradigm is certainly due to its ease of learning and use. Major search engines on the Internet, such as Google and others, use this paradigm for this reason. The result of the search is then a set of documents sorted by their relevance. The weakness of this approach lies in the inherent complexity of correctly interpreting the information need, as very few clues (query terms) are provided.
- **Structured Search:** In this paradigm, the information need is expressed in the form of a formal query, and the data is available in structured form. Examples include the use of relational databases with SQL as the query language, or triplestores and SPARQL as the query language. The strength of this approach is that it gives users a formalism to express their information needs very precisely. The difficulty lies in its application, as users are required to be familiar with the query language and able to use it. Furthermore, users must also know the structure of the data set, as only on this basis can queries be constructed.
- **Natural Language Search:** This paradigm is very similar to keyword search. Here, too, natural language queries are expected, which are structured much like those in keyword search. Typical queries in this paradigm start with: who, what, where, when, why, or how. In contrast to keyword search, the IRS in this case tries to first understand such a query and then provide a correct answer. For a query like "Who won an Oscar at the last awards ceremony?" a list of actor names would be expected as the answer. The advantage of this paradigm is again its natural usability for humans. The challenge for the IRS is to correctly interpret the query and resolve ambiguities.

> Semantics can lead to better results at various levels.

In the following, this subsection briefly discusses keyword search in combined data. After that, only topics related to keyword search in knowledge bases, especially in the specific form of ontology-based IR, will be addressed.

4.2.1   Keyword Search in Combined Data

A semantic search engine that performs keyword search in knowledge bases is Glimmer [33]. Glimmer is a search engine for RDF data and was released in 2013 by Peter Mika, Roi Blanco, and Sebastiano Vigna from Yahoo! Labs (Open Source). Their method was already presented in 2011 at the International Semantic Web Challenge (ISWC). The application is based on the Apache Hadoop framework, which is very popular in Big Data and implements the MapReduce method from Subsection 3.6.6 and achieves very effective and easily realizable scalability. This allows the highly computationally intensive indexing of RDF data to be performed in a distributed computer cluster, which significantly reduces runtimes. Furthermore, the application is based on the Java framework Managing Gigabytes for Java (MG4J) [34]. On the website, it is described as: "highly customisable, high-performance, full-fledged search engine providing state-of-the-art features (such as BM25/BM25F scoring) and new research algorithms." Based on the publications on the MG4J website, it is evident that an application based on the framework has already been presented at various IR and AI conferences.

For sorting relevant documents, Glimmer uses a modification of the ranking algorithm BM25F, which in turn extends the very popular BM25 algorithm in IR to enable ranking of structured documents with semantic annotation. The approach is similar to that of Apache Lucene (see Subsection 3.3.5.1), since the structured documents have fields (title, keywords, text, etc.) to which a specific "boost factor" can be assigned. This allows weighting of the fields for calculating the relevance of the documents.

Glimmer was demonstrated at the ISWC on a dataset of 750 million triples. It supports searches by types, combinations of type and relation, and various Boolean combinations. There is also the possibility to navigate through the ontology classes in a tree structure (so-called "Ontology Browsing"). The source code was published, among other places, on Github, including a small sample dataset (about 30 triples), various shell scripts for executing the necessary steps for indexing, and a Java web archive for deployment on an application server for browser-based search. Additionally, there is a guide on how to set up the application.

4.2.2   Keyword Search in Knowledge Bases

The challenge in keyword search in knowledge bases is, in general, to translate the inherently vague information needs of users into a formal query language such as SQL or SPARQL. The main motivation for using keywords is the simplified usability for users: there is no need to learn and use a formal language, and the structure of the data set does not need to be known. The core challenge here is first to map the terms of a query to the content and structure of the knowledge base and then to construct an appropriate formal query from the information obtained. The initial restriction is that the knowledge base to be searched can also be conceived of or represented as a graph. For the content of a knowledge base, this is directly the case. For the content of other forms, such as a relational database, this is at least not obviously the case, but can be established by considering database keys. In this subsection, however, we only consider search in knowledge bases. In practical terms, inverted indexes can be built for this mapping requirement. The indexes enable fast assignment of query terms to the nodes of the graph. However, this approach can also result in ambiguous assignments, i.e., assignments where an element of the query is mapped to more than one node of the graph.

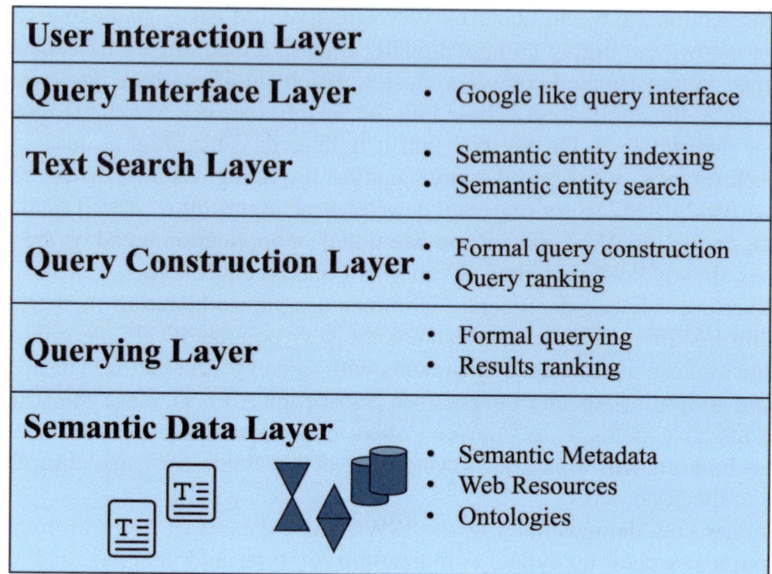

Fig. 4.5 Architecture of SemSearch [35]

A semantic search engine that performs keyword search in knowledge bases is SemSearch [35]. SemSearch consists of a five-layer system architecture from query reception to data storage (see Fig. 4.5). When processing a query, the following 6 steps are executed (see Fig. 4.6):

1. Queries are received in the form of keywords.
2. Keywords are mapped to the names of the nodes in the graph.
3. Possible formal queries are constructed.
4. If multiple queries can be formed, a relevance ranking or selection must take place.
5. The formal query is executed and a set of results is returned.
6. The found elements are sorted by their significance to the query.

4.2.3   Ontology Based IR

So-called ontology based IR is a type of semantic search that implements the keyword-search-in-knowledge-base paradigm in a narrowly defined way. Therefore, this subsection first defines exactly what is meant by ontology based IR. To illustrate the use of the approach, several methods that practically implement the approach are discussed. The following subsections then introduce the calculation of semantic similarity between sets of elements of an ontology. These

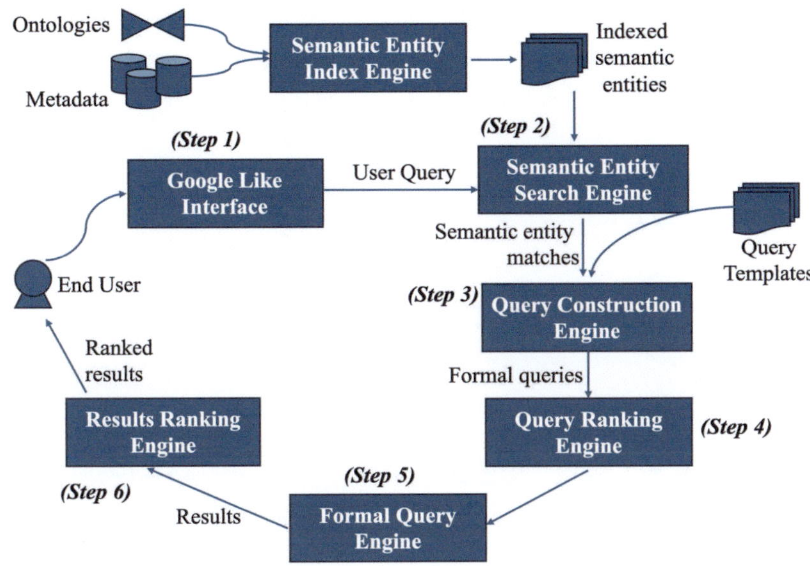

Fig. 4.6 Process of a semantic search [35]

calculations can be used within ontology based IR to derive the relevance between the query (represented by a set of elements of an ontology) and a document (also represented by a set of elements of an ontology). The explanations in this subsection follow Tran [36]. Here, ontology based IR is defined by four characteristics $\{D_O, Q, F_O, R(g_i, d_j)\}$ as follows:

- D_O is the ontology-based model in which a resource is represented by a set of elements $o \in O$ from the ontology. For this, we assume a function $f_T : D \rightarrow D_O$ that transforms a resource $d \in D$ into its ontology-based representation D_O.
- Q_O is a set of elements that represents the user's information need. The elements in Q_O correspond to the ontology elements in O. This enables the representation of Q_O in an ontology-based representation Q'_O.
- F_O is an ontology-based framework in which resources and queries are represented as ontology elements. An operation for consequence analysis checks whether the ontology-based representation of the resource is a consequence of the ontology-based representation of the information need.
- $R_O(q_i, d_i)$ is a ranking function.

The implementation of corresponding methods can be found, among others, in Vallet [37]. The method by Vallet is illustrated in Fig. 4.7. To calculate the RSV, Q_O and D_O are transformed into a vector representation. For this requirement, weights are calculated for each element of the query and a document, which are intended to reflect the importance for the document. For the calculation of the

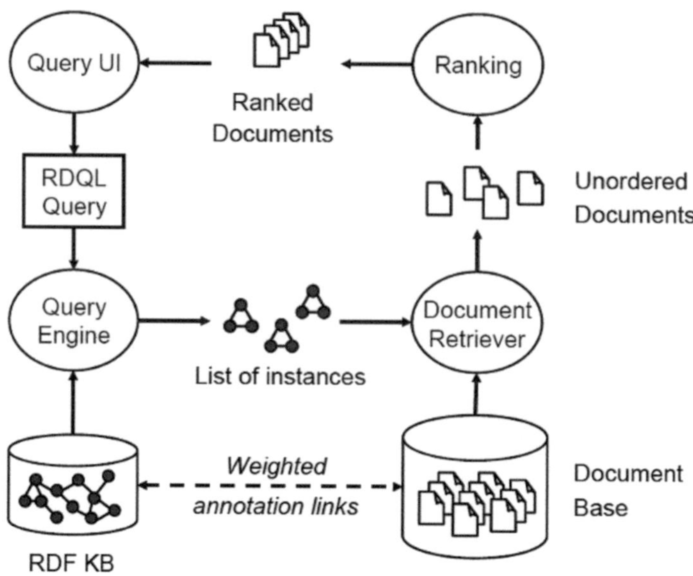

Fig. 4.7 Keyword Search in Knowledge Bases according to Vallet [37]

weights, an adaptation of TF/IDF is used. In this way, the dimensions of a vector are assigned weights. The RSV corresponds to the similarity of the query vector to a document vector. The similarity is calculated using the cosine. The method presented by Vallet overall shows a strong analogy to the classic VRM.

For the calculation of the RSV, i.e., the calculation of $R_O(q_i, d_i)$, a variety of possible calculations can in principle be used. The following chapter introduces methods that can be applied to calculate the similarities between two classes or two sets of classes.

4.2.3.1 Calculation of Semantic Similarity

The calculation of relevance within ontology based IR is based on the calculation of similarity between sets of elements of an ontology. In the methods presented so far, an approach based on the VRM was chosen for this requirement. In this subsection, alternatives to the transformation into a vector representation are discussed, which are implemented by calculating semantic similarity measures. Here, the relevance relationship between the query and a document (both represented by a set of elements of an ontology, see also Fig. 4.8) is derived by calculating the semantic similarity of two sets. The structure and some content of the following section are particularly based on the book by Harispe [38], which is generally referenced here.

For a fundamental understanding, it is important to emphasize that a "semantic measure" is only an umbrella term for calculations. Among other things, a distinction is made between the calculation of similarity of nodes in a graph and the

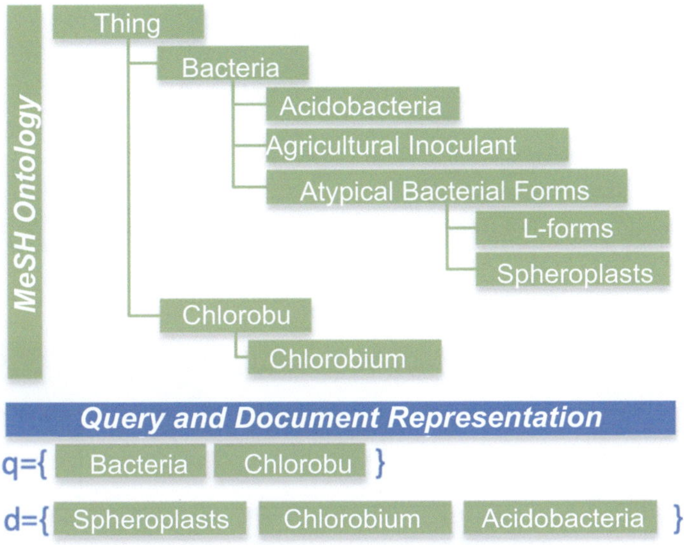

Fig. 4.8 Query and document representations using the elements of an ontology

calculation of relatedness of nodes. This difference should be briefly explained: Similarity refers to the closeness of two concepts in a parent-child relationship (an is-a relation). For example, in the MeSH thesaurus, the concepts Acidobacteria (DUI: D061271) and Spheroplasts (DUI: D013104) are children of the concept Bacteria (DUI: D001419) and thus are in an is-a relation. The closer the two concepts are in this hierarchy, the more similar they are considered. Relatedness, on the other hand, refers to a much more general association than that expressed by an is-a relationship. For example, Acidobacteria (DUI: D061271) is related to Bacterial Infections (DUI: D001424), since a bacterial infection is caused by bacteria. However, the two concepts are not in an is-a relation to each other. In general, it can be said that all similar nodes are also related, but not all related nodes are also similar.

In addition to similarity and relatedness, other terms are also used in the literature. The following list shows these (see Fig. 4.9):

- **Semantic relatedness** is sometimes also referred to as proximity, closeness, or nearness and corresponds to the concept introduced above.
- **Semantic similarity** has already been introduced. Sometimes the term taxonomical semantic similarity is used to express that only taxonomic elements are exploited here.
- **Semantic distance** generally refers to the inverse of semantic relatedness.
- **Semantic dissimilarity** corresponds to the inverse of semantic similarity.
- **Taxonomic distance** corresponds to the definition of semantic dissimilarity.

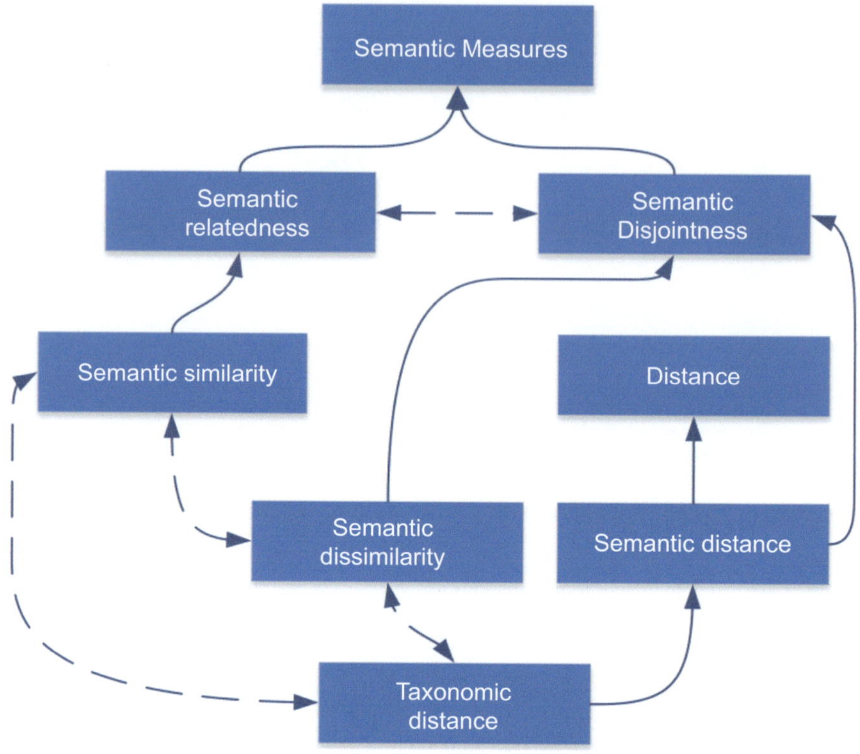

Fig. 4.9 Relationships of semantic similarity measures

For a precise introduction to semantic measures, it is also necessary to address what is meant mathematically by the terms distance and similarity. For this purpose, the necessary definitions are given in this subsection. Since general mathematical relationships are introduced here, we now (in contrast to the previous subsections) speak of unspecified elements of a set.

First, D denotes the set of all elements of a given application domain, and V the set on which a total order $(V, \leq)$ is defined. It holds that $\min_V, \max_V \in V$ and $\forall v \in V : \min_V \leq v, \max_V \geq v$. On this basis, a distance can be defined as a function $dist : D \times D \to V$, where for all $x, y \in D$ the following holds:

- Non-negativity: $dist(x, y) \geq \min_V$ and $\min_V = 0$
- Symmetry: $dist(x, y) = dist(y, x)$
- Reflexivity: $dist(x, x) = \min_V$ and $f \forall y \in D \wedge y \neq x : dist(x, x) < dist(x, y)$

The similarity of two elements from D is described in the form of a function as $sim : D \times D \to V$, where all $x, y \in D$ are non-negative, and symmetry and

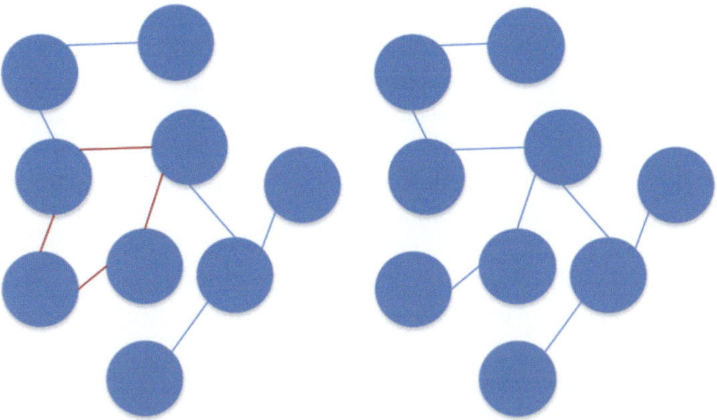

Fig. 4.10 Cyclic and acyclic graph

reflexivity are maintained. For a normalized similarity or distance function, it is also required that the result of the calculation must lie in the interval [0, 1].

4.2.3.2 Graph-Based Semantic Measures

Graphs can take different forms. A property to be highlighted here is the distinction between cyclic and acyclic graphs. Simply put, a cyclic graph is a graph that contains a cycle, and an acyclic graph is a graph that does not allow such a cycle (see Fig. 4.10).

The introduction of methods that are applied to acyclic graphs can be kept brief. This is because all methods that work in a cyclic graph can also be applied in a cyclic graph that has been reduced to an acyclic graph. One form of reduction to ensure that no cycles are present would be, for example, to consider only is-a relations of a graph in the calculation. Most methods for calculating semantic similarity are limited to this form of similarity calculation.

The calculation on cyclic graphs can be divided into two classes: graph traversal methods and methods that exploit certain properties of the graph. Both classes will be discussed below. The graph traversal methods can generally be characterized by the following distinctions:

- **Shortest path:** In general, the length of the connection between two nodes of the graph is evaluated. The fewer edges between two nodes, the more similar the nodes are. Applied to relations in a thesaurus or ontology, this also means that every relation can be included in the calculation. This property is critical with regard to semantic expressiveness, since the relations between classes in particular convey specific semantics.
- **Random walks:** are based on so-called Markov chains. Generally speaking, Markov chains model a stochastic process that calculates the occurrence of a

particular event. Applied to a graph that expresses semantics, this means calculating the probability of reaching the next concept via a relation. Various proposals have been made to use random walk for calculating semantic similarity—including measuring the average number of steps taken to get from u to v, or calculating the commute time, which is the sum of the steps from u to v and back. In this application, the more paths connect u and v, the shorter the commute time and the greater the semantic similarity.

In addition to the graph traversal methods, there are methods that exploit certain properties of a graph. For example, ontologies expressed in Semantic Web languages have properties that have a specific data type (e.g., string or integer) and assign a specific value to instances of a class. Calculations that exploit the properties of graphs are based on comparing these properties and the instances of classes.

4.2.3.3 Calculation of Pairwise Semantic Similarity

To introduce the measures that perform pairwise comparisons, an analogous subdivision to the previous chapter is made—on the one hand, into calculations based on the structural characteristics of the graph, and on the other hand, into calculations based on features. However, it should generally be noted that the distinction cannot always be made sharply, as often features of the structural form of the graph are also exploited in the analysis.

Many of the methods that implement **structure-based approaches** are based on traversing a graph while considering certain additional properties (extensions). A very simple distance measure is that of Rada [39]. It calculates the distance between two classes based on the shortest path (SP) via an is-a relationship:

$$dist_{\text{Rada}}(u.v) = sp(u, isA, v) \tag{4.1}$$

From this definition of a distance measure, a definition of a similarity measure can easily be derived:

$$sim_{\text{Rada}}(u, v) = \frac{1}{dist_{\text{Rada}}(u, v) + 1} \tag{4.2}$$

It is clear in these calculations that u and v have a least common ancestor (LCA). The LCA can also be considered as an indicator of the expressiveness of a connection. An example of this is the calculation according to Resnik [40] (where G_T denotes the graph of a taxonomy):

$$sim_{\text{Resnik-eb}}(u, v) = 2 \cdot depth(G_T) - sp(u, isA, LCA(u, v)) - sp(v, isA, LCA(u, v)) \tag{4.3}$$

Another measure in this class is that of Slimani [41]. In his calculation, Slimani introduces a parameter λ, which ensures that neighboring classes are not more similar than classes that are in the same taxonomic hierarchy:

$$sim_{tbk}(u, v) = sim_{WP}(u, v) \times pf(u, v) \tag{4.4}$$

where

$$pf(u, v) = (1 - \lambda)(\min(depth(u), depth(v)) - \max_{depth}) \tag{4.5}$$

$$+ \lambda(depth(u) + depth(v) + 1)^{-1} \tag{4.6}$$

A disadvantage of these methods is that traversing the graph is quite expensive in terms of memory and runtime behavior. However, graphs can be mapped into mathematical spaces via their adjacency matrix, which—depending on the structure of the graph—can enable significantly more efficient calculations.

Feature-based approaches use certain properties to calculate similarities. As a representative example, the framework by Ehrig [42] for calculating similarities between instances of an ontology will be introduced. It should be briefly mentioned again that values assigned to a class are referred to as instances. Instances are connected to other instances via relations. Relations have a defined domain and range.

Ontologies correspond to the conceptualization of an application domain and represent a model of relationships that goes beyond the syntactic level. The framework is based on the assumption that, for a similarity comparison of instances of an ontology, the reference to the real world should be included in the calculation. Ehrig refers to this as the external context. For example, the purchasing behavior of customers on an online sales platform can be used to infer whether two book offers are the same (assuming that similar entities also exhibit similar usage patterns). This context is used to configure similarity functions that interact on three levels. These are, specifically, the data level, the ontology level, and the context level (see Fig. 4.11).

The data level is the model layer in which calculations are performed on simple or complex data types such as strings or integers. Here, calculations such as the Hamming distance [43], the Levenshtein distance [44], or the Jaro measure [45] are applied, among others. At the ontology level, the structure of the ontology is included in the similarity calculation. This level includes, among others, measures that take into account similarity calculations over the taxonomic structure of the ontology. At the context level, by contrast, two instances are compared based on how they are used in an external context.

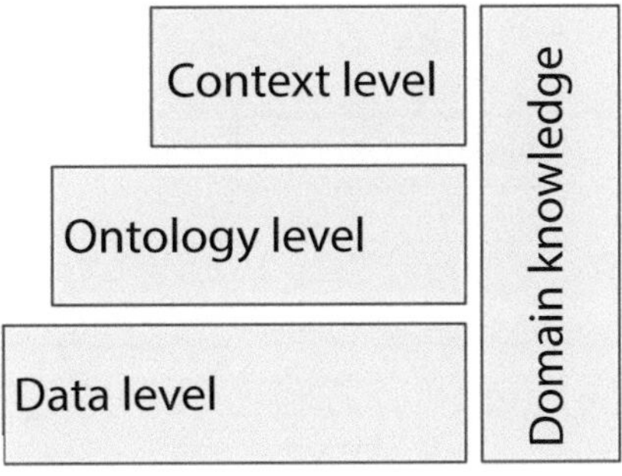

Fig. 4.11 Layer model according to Ehring

One possible context manifestation is the application context. The application context corresponds to the external knowledge of domain experts about the application and the environment of a modeled instance of the ontology. For example, if the similarity of two person instances in a scientific domain is to be calculated, a possible application context would be to consider the fact that authors of scientific publications are connected to other authors via co-authorship. In this case, two person instances would be more similar the more common co-authors they have. Another example is the intersection of shared publications between two authors. The overall result of these individual considerations is aggregated using a so-called fusion function, which sums the individual results and maps them to a similarity value. A concretization of Ehrig's framework with regard to the formalization of the application context is described in the work of Albertoni [46].

Albertoni describes a framework for calculating semantic similarity between instances of a knowledge base, which builds on Ehrig's model. In particular, Albertoni's work includes a formalization of the application context, which will be introduced below. The first building block for formalization is the definition of a set of ontology elements (sets of attributes A and relations R of instances) that are part of the calculation and are combined into a sequence. Given a set X of elements of the ontology, a sequence s of elements from X of length $n \in N$ is defined as follows:

$$S_X^n = \{s|s : [1, n] \to X\} \tag{4.7}$$

Based on this function definition, the so-called recursive path is defined. A recursive path p of length i is a sequence of relations that can be reached from an ontology concept C via connected relations R. According to the ontology in Fig. 4.12, a

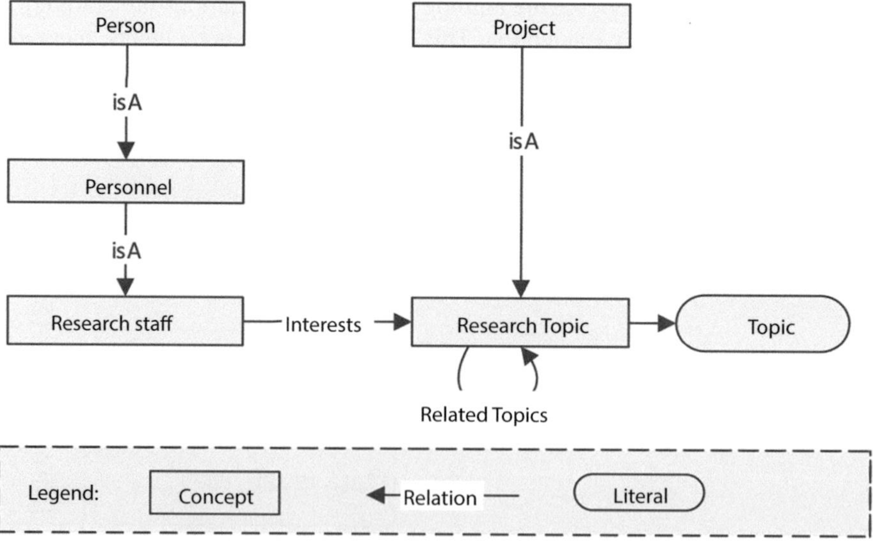

Fig. 4.12 Excerpt from a sample ontology

possible recursive path would be: [Research Staff, Interest, Related Topics]. Along this path, two people could be compared with respect to their research topics. To express this comparison as a numerical value, a function is needed that computes the elements reached via this path and determines their similarity.

For the calculation of the elements selected via a recursive path, Albertoni et al. [46] introduce three simple operations. The operation *count* calculates the cardinality of the elements (e.g., if two author instances have the same number of publications, their similarity increases). The operation *inter* computes the intersection of the elements (the larger the intersection between the publications of two person instances, the more similar their interests). The operation *simil* calculates the similarity between attributes directly or via attributes that are reachable through relations. The set of operations defined by Albertoni is thus established as the set $L = \{count, inter, simil\}$. Overall, the application context according to Albertoni consists of the interplay of these introduced elements and calculations and is defined via the mapping function AC:

$$AC : P \rightarrow (2^{A \times L}) \times (2^{R \times L}) \tag{4.8}$$

The variable P in this definition corresponds to a set of recursive paths.

4.2.3.4 Semantic Similarity Between Sets of Nodes

When comparing sets of nodes in a graph, two basic groups of approaches can be distinguished. Direct comparison refers to approaches that access the information present in the graph directly. Indirect methods use pairwise comparisons, the results of which are then aggregated in the form of a Cartesian product or similar.

- **Direct comparison of concept groups:** In principle, the calculation of concept groups is a generalization of pairwise comparisons. Set operations can generally be used, or a representation in the form of vectors, to then perform a distance measurement using the cosine measure, analogous to the VRM.
- **Structure-based approaches:** Here, the structure of the graph is directly exploited, for example by forming the set of ancestors of all classes. The semantic similarity of the sets then results from evaluating the longest path.
- **Feature-based approaches:** Analogous to the previous subsection, certain properties of the graph are also used here to calculate similarities, i.e., only certain types of relationships are used or weighted more heavily.
- **Indirect comparison of concept groups:** These methods are based on the application of the pairwise methods from the previous subsection. In principle, the similarity of the pairs can be calculated by forming a product or sum, followed by normalization of the possible comparisons. For this, a so-called aggregation function is used

4.2.4 Semantic Measure Library

A freely available Java library in this context, which can be used to calculate semantic measures, is the Semantic Measures Library (SML) [47]. The SML is available both as a library for integration into your own Java projects and as a toolkit for direct use on the command line, and provides a programmatic interface for calculating the computations introduced in the previous subsection. The SML can be used with various knowledge representation systems, including RDF(S) graphs, OWL ontologies, WordNet, MeSH, Gene Ontology (OBO ontologies), SNOMED-CT. The SML supports only graph-based measures for calculating similarities.

A fundamental distinction here is between pairwise and groupwise measures: While pairwise measures compare two individual concepts—for example, D016779 (Malaria, Cerebral) and D010272 (Parasitic Diseases)—the groupwise measures can be applied to two groups or sets of classes. These can be performed directly with the groupwise direct measures. However, the pairwise measures do not have to be excluded, as they can also be used in conjunction with the groupwise indirect measures. Here, pairwise comparisons of the individual concepts are performed, and depending on the groupwise indirect measure, the maximum, a mean value, or similar is then determined. The pairwise measures operate either edge-based (these are essentially shortest-path algorithms) or node-based. The latter are in part IC-based, i.e., they additionally require an information content (IC)—this is used in node-based methods to determine the value or specificity of terms, i.e., it measures how informative a term is.

The operation of the SML API with respect to MeSH will be examined in more detail here. The examination is divided into the processing steps: creating the graph, loading it into the API, and calculating the similarity. One challenge in calculating the similarity of MeSH elements is the creation of a graph on which the calculation is to take place. The challenge lies in the fact that MeSH descriptors are classified into different tree structures (across the 16 categories, such as Diseases and Chemical and Drugs). It is important to note that these MeSH tree structures do not have a root node by default. However, this is essential for the calculation of similarity and is therefore artificially introduced (see Fig. 4.13). Next,

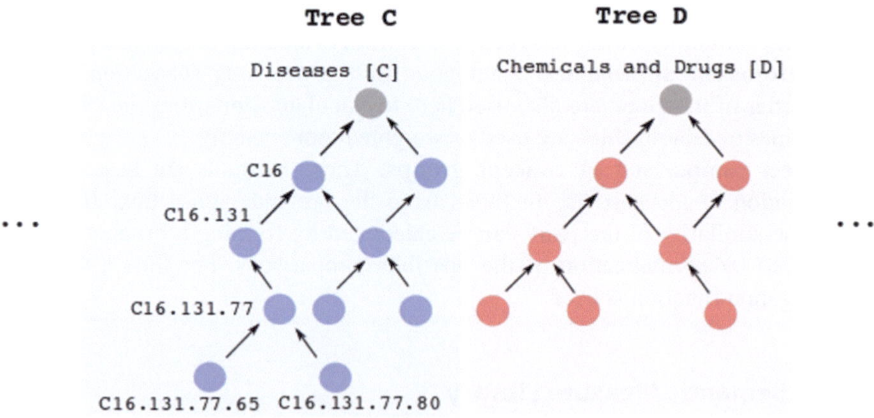

Fig. 4.13 MeSH tree structures

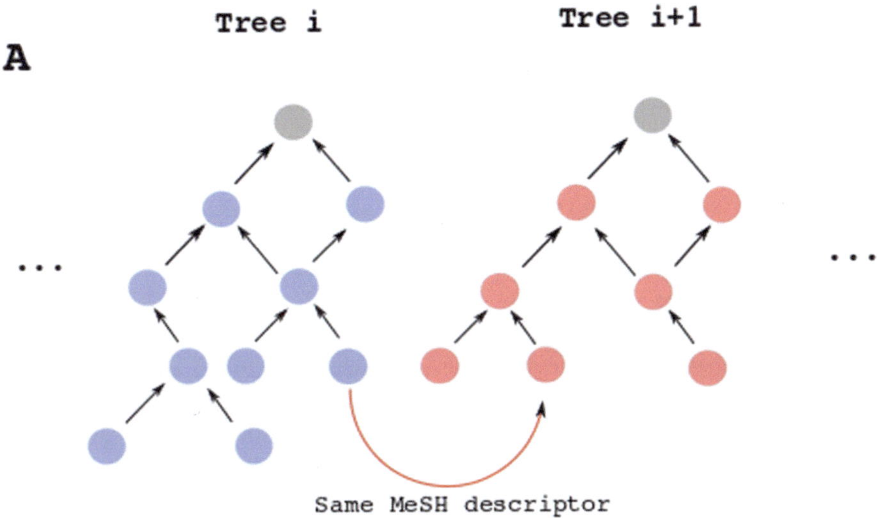

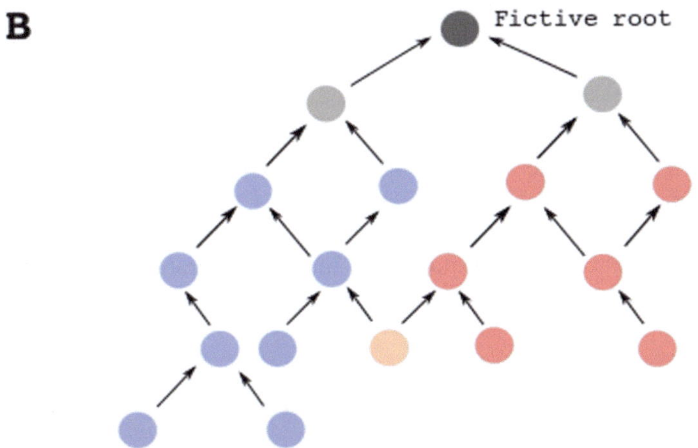

Fig. 4.14 Merged MeSH tree structures

it is checked whether the two trees to be merged contain the same descriptors. If so, the merging is performed in the next step and a root node is again introduced for the merged tree (see Fig. 4.14).

The calculations introduced in the previous chapter are based on processing a graph. However, the tree resulting from the merging must be a rooted directed acyclic graph (rDAG) for the calculation of similarities. The construction of a MeSH graph can, for example, be organized hierarchically via broader-narrower relationships (e.g., using SKOS). However, it must be ensured that no transitive connections are established and thus no inferences of the form "if 'animals' is broader than 'mammals' and 'mammals' is broader than 'cats', then 'animals' is broader than 'cats'" (see Fig. 4.13) can be made.

However, when merging graphs, cycles may occur. To resolve such conflicts, the SML API removes the connections that lead to acyclic graphs. This approach does, to some extent, distort the connections originally specified by MeSH. However, this is accepted in view of usability.

4.3   Summary

In this comprehensive chapter, we have introduced you to the models and concepts of information retrieval. Exact-match methods return results only when they exactly match the query, while best-match methods introduce a relevance judgment and deliver the best possible results, even if they are not exact. You have learned about various optimization options, such as relevance feedback, and seen some libraries for concrete implementation.

Another important subsection of this chapter deals with the area of data organization, which includes persistent identifiers, content distribution, and thus also distributed information retrieval and long-term archiving.

Finally, the area of semantics was brought together with information retrieval. Compared to the established field of full-text document search, semantic search is still a relatively young discipline with many open questions and challenges. The acceptance of semantic search has been particularly fostered by the emergence of the Semantic Web. The Semantic Web provides new mechanisms that enable the use of explicit and machine-readable semantics in search. Commercially, semantic search, as introduced in this chapter, has not yet achieved major successes. This may be due, not least, to the very computationally intensive operations required to calculate RSVs (see Fig. 4.15).

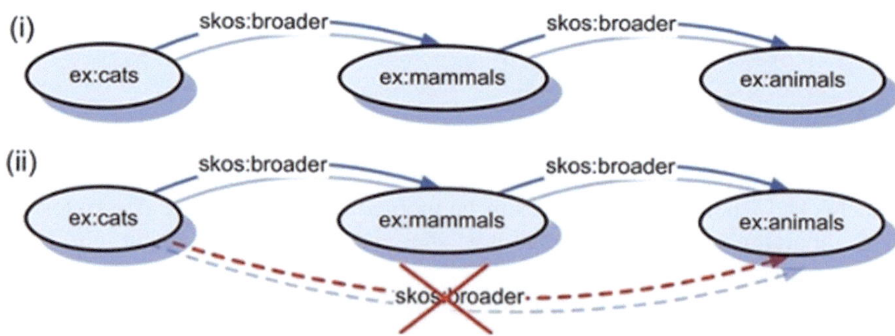

Fig. 4.15 Excerpt from a SKOS graph

4.4   Self-Assessment Questions

4.1—Long-Term Archiving

Which types of metadata are relevant for long-term archiving?

4.2—Long-Term Archiving

What role does the "Intellectual Entity" play in the PREMIS data model?

4.3—Long-Term Archiving

What is a "SIP" in the OAIS reference model?

4.4—Semantics

What characterizes the random walk approach?

References

1. W3C. "SPARQL Query Language for RDF." (Aug. 2013), [Online]. Available: https://www.w3.org/TR/sparql11-overview/, Download: 14.12.2021.
2. I. Maglogiannis, C. Doukas, G. Kormentzas, and T. Pliakas, "Wavelet-based compression with roi coding support for mobile access to dicom images over heterogeneous radio networks," *Information Technology in Biomedicine, IEEE Transactions on*, vol. 13, pp. 458–466, 2009. https://doi.org/10.1109/TITB.2008.903527.
3. C. Webb, "Digital preservation: A many-layered thing – experience at the national library of australia," in *Preserving our Digital Heritage*, P. Conway and R. Sproull, Eds., Council on Library and Information Resources (CLIR), 2001, pp. 63–71. [Online]. Available: https://www.clir.org/pubs/reports/pub107/webb/.
4. S. Rohde-Enslin, *Nicht von dauer. kleiner ratgeber für die bewahrung digitaler daten in museen*, 2004. [Online]. Available: https://www.langzeitarchivierung.de/downloads/ratg/ratg01.pdf.
5. P. Caplan, *Understanding premis*, 2008. [Online]. Available: https://www.loc.gov/standards/premis/understanding-premis.pdf.
6. National Information Standards Organization (NISO), *Ansi/niso z39.87-2006 (r2017): Data dictionary – technical metadata for digital still images*, https://www.niso.org/publications/ansiniso-z3987-2006-r2017-data-dictionary-technical-metadata-digital-still-images, American National Standards Institute (ANSI) approved standard, 2017.
7. PREMIS Editorial Committee, *Premis data dictionary for preservation metadata, version 3.0*, https://www.loc.gov/standards/premis/v3/, Library of Congress, 2015.
8. W. C. Lee and E. A. Fox, "Experimental comparison of schemes for interpreting boolean queries," 1988.
9. Library of Congress, *Metadata encoding and transmission standard (mets)*, https://www.loc.gov/standards/mets/, Accessed: 2025-05-19, 2025.
10. Consultative Committee for Space Data Systems (CCSDS), *Reference Model for an Open Archival Information System (OAIS)*. CCSDS Secretariat, Space Communication and Navigation Office, Washington, D.C., 2012.

11. V. Slavina and Y. Soldatkina, "Media culture as the information age phenomenon," *RUDN Journal of Studies in Literature and Journalism*, vol. 26, pp. 286–293, 2021. https://doi.org/10.22363/2312-9220-2021-26-2-286-293.

12. Dublin Core Metadata Initiative (DCMI). "The dublin core metadata initiative (dcmi)." https://dublincore.org.

13. T. Steinke, *Langzeitarchivierungsmetadaten für elektronische ressourcen (lmer), version 1.2*, https://d-nb.info/97511087x/34, Deutsche Nationalbibliothek, 2005.

14. T. Steinke, *Universelles objektformat: Ein archiv- und austauschformat für digitale objekte*, http://kopal.langzeitarchivierung.de/downloads/kopal_Universelles_Objektformat.pdf, Deutsche Nationalbibliothek, 2006.

15. Library of Congress, *Marc 21 format for bibliographic data*, https://www.loc.gov/marc/bibliographic/, Update No. 39 (December 2024), 2024.

16. Library of Congress, *Metadata object description schema (mods)*, https://www.loc.gov/standards/mods/, Version 3.8, 2022.

17. Library of Congress, *Metadata authority description schema (mads)*, https://www.loc.gov/standards/mads/, Version 2.1, 2016.

18. P. W. Group, *Implementing preservation repositories for digital materials: Current practice and emerging trends in the cultural heritage community*, https://lccn.loc.gov/2004541476, OCLC Online Computer Library, Inc., 2004.

19. L. Meyer and J. Gertz, *Rlg guidelines for microfilming to support digitization*, https://www.oclc.org/content/dam/research/publications/library/Pres_Micro_Supplement.pdf, Research Libraries Group, 2003.

20. C. A. Lee, "Open archival information system (oais) reference model," *Encyclopedia of Library and Information Sciences*, 2005. [Online]. Available: https://ils.unc.edu/callee/p4020-lee.pdf.

21. T. van der Werf-Davelaar, "Long-term preservation of electronic publications: The nedlib project," *D-Lib Magazine*, vol. 5, no. 9, 1999. [Online]. Available: https://dl.acm.org/doi/abs/10.5555/865579.

22. K. Russell, "Digital preservation and the cedars project," *Microform & Digitization Review*, vol. 27, no. 4, pp. 145–151, 1998. https://doi.org/10.1515/mfir.1998.27.4.145. [Online]. Available: https://www.degruyter.com/document/doi/10.1515/mfir.1998.27.4.145/html.

23. Consultative Committee for Space Data Systems (CCSDS), "Reference model for an open archival information system (oais)," CCSDS, Tech. Rep. CCSDS 650.0-B-1, 2002, Blue Book. [Online]. Available: http://public.ccsds.org/publications/archive/650x0b1.pdf.

24. O. A. Initiative, *The open archives initiative protocol for metadata harvesting version 2.0*, https://www.openarchives.org/OAI/openarchivesprotocol.html, Accessed: 2025-05-19, 2002.

25. D. Nationalbibliothek and N. S. und Universitätsbibliothek Göttingen, *Kopal – kooperativer aufbau eines langzeitarchivs digitaler informationen*, https://kopal.langzeitarchivierung.de/index.php.en, Accessed: 2025-05-19, 2007.

26. E. L. Group, *Digitool – digital asset management system*, https://www.exlibrisgroup.com/products/digitool-digital-asset-management/, Accessed: 2025-05-19.

27. I. D. GmbH, *Digital information and archiving system (dias)*, https://www.dnb.de/EN/Professionell/Erhalten/LZA-System/lza-system_node.html, Accessed: 2025-05-19, 2004.

28. S. Payette and C. Lagoze, "Flexible and extensible digital object and repository architecture (fedora)," *International Journal on Digital Libraries*, vol. 5, no. 4, pp. 319–328, 2005. https://doi.org/10.1007/s00799-004-0102-5.

29. U. of Southampton, *Eprints repository software*, https://www.eprints.org/, Accessed: 2025-05-19.

30. M. Community, *Mycore – open source framework for digital repositories*, https://www.mycore.de/, Accessed: 2025-05-19.

31. R. Tansley, M. Kimpton, L. Lannom, and M. Stuve, "The dspace institutional digital repository system: Current functionality," *D-Lib Magazine*, vol. 9, no. 1, 2003. https://doi.org/10.1045/january2003-tansley.

32. H. Bast, B. Buchhold, and E. Haussmann, "Semantic search on text and knowledge bases," *Foundations and Trends® in Information Retrieval*, vol. 10, no. 2–3, pp. 119–271, 2016, ISSN: 1554-0669. https://doi.org/10.1561/1500000032. [Online]. Available: http://dx.doi.org/10.1561/1500000032.

33. K. Idehen et al., *Re: Glimmer – semantic search engine [was: Callimachus, semantic data wiki platform]*, https://lists.w3.org/Archives/Public/semantic-web/2013Jun/0157.html, Message posted on W3C Semantic Web mailing list, 2013.

34. P. Boldi and S. Vigna, "MG4J at TREC 2005," in *The Fourteenth Text REtrieval Conference (TREC 2005) Proceedings*, E. M. Voorhees and L. P. Buckland, Eds., ser. Special Papers, http://mg4j.di.unimi.it/, NIST, 2005.

35. F. von Henke, T. Liebig, and O. Noppens, "Semsearch – die benutzergeführte semantische suche," Institut für Künstliche Intelligenz, Universität Ulm, Tech. Rep., 2007, Accessed: 2025-05-19.

36. T. Tran, S. Bloehdorn, P. Cimiano, and P. Haase, "Expressive resource descriptions for ontology-based information retrieval," 2007.

37. D. Vallet, M. Fernández, and P. Castells, "An ontology-based information retrieval model," in *The Semantic Web: Research and Applications*, A. Gómez-Pérez and J. Euzenat, Eds., Berlin, Heidelberg: Springer Berlin Heidelberg, 2005, pp. 455–470, ISBN: 978-3-540-31547-6.

38. S. Harispe, S. Ranwez, S. Janaqi, and J. Montmain, *Semantic Similarity from Natural Language and Ontology Analysis*. May 2015, vol. 8. https://doi.org/10.2200/S00639ED1V01Y201504HLT027.

39. R. Rada, H. Mili, E. Bicknell, and M. Blettner, "Development and application of a metric on semantic nets," *IEEE Transactions on Systems, Man, and Cybernetics*, vol. 19, no. 1, pp. 17–30, 1989. https://doi.org/10.1109/21.24528. [Online]. Available: https://ieeexplore.ieee.org/document/24528.

40. P. Resnik, "Using information content to evaluate semantic similarity in a taxonomy," in *Proceedings of the 14th International Joint Conference on Artificial Intelligence (IJCAI)*, San Francisco, CA, USA: Morgan Kaufmann, 1995, pp. 448–453. [Online]. Available: https://doi.org/10.5555/1625855.1625914.

41. T. Slimani, B. B. Yaghlane, and K. Mellouli, "A new similarity measure based on edge counting," in *Proceedings of the World Academy of Science, Engineering and Technology*, vol. 20, 2006, pp. 34–38. [Online]. Available: https://www.researchgate.net/publication/228723587_A_New_Similarity_Measure_based_on_Edge_Counting.

42. M. Ehrig, P. Haase, M. Hefke, and N. Stojanovic, "Similarity for ontologies – a comprehensive framework.," Jan. 2005, pp. 1509–1518.

43. R. W. Hamming, "Error detecting and error correcting codes," *Bell System Technical Journal*, vol. 29, no. 2, pp. 147–160, 1950. https://doi.org/10.1002/j.1538-7305.1950.tb00463.x.

44. V. I. Levenshtein, "Binary codes capable of correcting deletions, insertions and reversals," *Soviet Physics Doklady*, vol. 10, p 707, 1966, Original article submitted January 2, 1965. [Online]. Available: https://www.bibsonomy.org/bibtex/55f7ad93fcb9ae3ed999afaa6e24937d.

45. M. A. Jaro, "Advances in record-linkage methodology as applied to matching the 1985 census of tampa, florida," *Journal of the American Statistical Association*, vol. 84, no. 406, pp. 414–420, 1989. https://doi.org/10.1080/01621459.1989.10478785.

46. R. Albertoni and M. De Martino, "Semantic similarity of ontology instances tailored on the application context," Oct. 2006, pp. 1020–1038, ISBN: 978-3-540-48287-1. https://doi.org/10.1007/11914853_66.

47. S. Harispe, S. Ranwez, S. Janaqi, and J. Montmain, "The semantic measures library and toolkit: Fast computation of semantic similarity and relatedness using biomedical ontologies," *Bioinformatics*, vol. 30, no. 5, pp. 740–742, 2014. https://doi.org/10.1093/bioinformatics/btt581. [Online]. Available: https://academic.oup.com/bioinformatics/article/30/5/740/245711.

Evaluation and Relevance Metrics

5

So far, the concept of relevance has been frequently used in this book without further elaboration on its precise meaning and definition. In fact, an in-depth discussion and definition of relevance is another cornerstone for explaining IR models and, in particular, for evaluating their application.

Let us first recall that, in the introduction to the information-seeking models in Subsection 1.4, the information need of users was described. The information need is presented as the driving force for using an IRS. The requirement for the IRS is to output documents relevant to the information need. The information need is usually vague; users of the IRS often cannot specify their need for information precisely. In Subsection 3.3, it was described that IR models can be divided into the classes Exact Match and Best Match. A significant difference between the IR models of both classes is that Best Match models calculate an RSV, which serves as the basis for sorting the output documents by their relevance to the query. In general, it is assumed that the better the IRS implements the sorting of documents ranked by RSV in descending order of relevance, the higher its effectiveness.

5.1 General Introduction

General definitions from well-known dictionaries define relevance as follows: "relation to the matter at hand" [22] or "importance in a particular context" [23]. To shed more light on these definitions and provide some substance, we first take a linguistic perspective, following Wilson and Sperber [7]. Building on the insights gained there, we then present works on this topic from the field of IR.

Wilson and Sperber jointly developed the so-called Relevance Theory. They base their view of relevance on the fundamental assertion that an essential component of communication lies in expressing and recognizing a particular intention. A simple yet fundamental model of communication that addresses this assertion is

© The Author(s), under exclusive license to Springer-Verlag GmbH, DE, part of
Springer Nature 2026
M. Hemmje and S. Wagenpfeil, *Multimedia Information Retrieval*,
https://doi.org/10.1007/978-3-662-73310-3_5

the so-called **coding model**: Here, a communicator encodes their intention into a signal, the signal is received by a recipient, and decoded. Another model, the so-called **inferential model**, states that the communicator provides evidence of their intention to convey a particular meaning, which is then inferred by the recipients based on the evidence. The listener of a message thus derives the meaning of the message based on evidence. The evidence is based on the principles of quality (truthfulness), quantity (informativeness), relation (relevance), and manner (clarity). Naturally, all these principles should best meet the listener's expectations. A central demand that Wilson and Sperber raise in Relevance Theory is that the expectation aroused by an utterance should be precise and predictable enough to convey the meaning of the communicator's signal to the listener. Based on these models, Relevance Theory states that any external or internal stimulus that serves as input for a cognitive process can be relevant to a person at a given time. An intuitive answer from Relevance Theory to the question of when such a stimulus can be relevant to a person is: when it connects with existing information and an inference can be made that is important to the person.

Communication is an integral part of information systems, especially in IRS. In IR, among others, the information-seeking models have already referred to communication between a sender and a receiver (e.g., the ASK model). Furthermore, Kuhlen also emphasizes the communicative character of information in his definition. A concise description of relevance in connection with IRS from a dictionary describes relevance as "the ability (as of an IR system) to retrieve material that satisfies the needs of the user" [24]. This rather simple view will now be considered in more detail. The aim is to clarify how relevance is understood in information science. The following discussion is based in particular on the publication [8] by Saracevic.

Saracevic notes that, in the state of science and technology, there are four thematic areas that repeatedly spark discussions in information science:

1. **Types of relevance:** Which environment or, more generally, which framework is suitable for defining relevance in order to investigate the many possible manifestations of relevance?
2. **Manifestations of relevance:** In what context does relevance manifest itself? Can manifestations of relevance be classified?
3. **Behavior of relevance:** What are the observable characteristics of relevance in a given context? How does this relate to information seeking?
4. **Effects of relevance:** How can relevance be used in work on the development of information systems, processes, and evaluations?

The following provides an overview of the first two thematic areas: types and manifestations of relevance. To begin, let us discuss the types of relevance, noting first that users of an IRS seek and acquire information in order to use it. This process is highly dynamic and is based on the use of a computer. The computer contains resources, such as multimedia objects, which in turn have their own cognitive structure and representation. IR is thus understood as an interaction between users and the computer via an interface. The intention is to address the affective (i.e., emotional) state of the users to enable effective use of the available information. IR interaction thus takes place on different levels.

- **Interaction with the interface**: Here, the interaction consists of a sequence of events. Users engage in a dialog with the computer, which is not limited to searching. Processes of understanding, browsing, or visualizing take place, as well as inferences about relevance. Computers, on the other hand, interact with users based on their own knowledge and on answers to questions, thus providing information. Here too, relevance plays a role and is determined by a process or algorithm.
- **Interaction with cognition:** Users interact with a given multimedia object (e.g., text, image, audio, video), interpret it, and try to understand its content. This cognitive interaction inherently involves deriving relevance with respect to a given requirement.
- **Interaction with the situation:** Users interact with the computer in view of a current challenge that has created an information need, which is formulated as a query. During the search, the information need is reinterpreted, new facets are introduced, and as a result, the query is reformulated.
- **Interaction with emotions**: Users often interact with certain intentions and associated emotions (e.g., urgency, dissatisfaction, or joy). The determination of relevance also depends on this level.

At this point, it should be emphasized again that virtually every human-computer interaction can be described by IR processes. Clicking the "Like" button can be realized via relevance feedback, clicking on a (predefined) link is an exact match retrieval (because a precisely specified target page is found and displayed), and selecting a detailed record from a hit list can be interpreted as query refinement. Once again, this consideration makes it clear that IR/MMIR is highly relevant in every area of computer science.

Relevance is influenced at different levels by the users of a system, but also by the developers of an involved algorithm, who also make certain assumptions about relevance, analogous to the principles from Relevance Theory at the beginning of this chapter. Both users and computers have a particular view of relevance, taking into account the various levels involved. This can also be seen as a system of relevances. Relevance in IR must therefore be regarded as a multidimensionally influenced metric that can appear in various forms. The question that arises at this level is whether there is a way to classify, generalize, or specify the different manifestations of relevance. Saracevic describes a consensus in the state of research regarding the following five levels [8]:

1. System or algorithmic relevance: The selection of a document by an IRS is made by comparing representations of the query and the documents based on an algorithm. The automated comparison is always based on certain assumptions underlying the comparison.

2. Topicality or subjective relevance: The connection between the query and a relevant document should be the same subject matter. Topicality or subjective relevance results from the sameness of the topic of the two representations.
3. Cognitive relevance or pertinence: The connection between the available knowledge of the users and the cognitive need determines cognitive relevance (e.g., information content, novelty, or quality of the information).
4. Situational relevance or utility: This consists of the connection between the situation, task, and the document found by the IRS. Here, factors such as usefulness for making a decision or reducing uncertainty play a role.
5. Motivational or affective relevance: This consists of the connection between the intention, goals, and motivation of users and the documents found. Satisfaction and success are among the criteria for deriving motivational or affective relevance.

This subdivision of the manifestations of relevance is also a suitable complement to the previously introduced levels of interaction. Here too, given the multidimensional nature of relevance, it seems a good approach to describe the manifestations in different, interacting levels. In summary, this discussion shows that relevance is a fundamental building block of all considerations in the IR process. Relevance is a multifaceted concept that can be viewed and defined from various perspectives. Simply put, relevance in IR is a metric by which the interaction of users with the IRS can be described and evaluated. Again, it is true that relevance can be an application- and user group-dependent concept, and a clear definition is needed for the evaluation of IR processes as to what "relevance" is supposed to mean in the context of the application, the user group, the data base, and the evaluation.

5.2   Effectiveness Analyses by Benchmarking

The question arises as to how relevance, as a metric for the quality of an IRS, can be practically calculated. In general, IR distinguishes between user-oriented evaluation and system-oriented evaluation. User-oriented evaluation involves users of IR methods in the evaluation in order to draw conclusions about the quality of an IR process based on their feedback. System-oriented evaluation does not involve users and focuses on assessing system performance (efficiency) and effectiveness. The following subsections deal with the system-oriented evaluation of IR processes to assess their effectiveness in the task of outputting ranked relevant documents for a query.

Effectiveness refers to the accuracy and completeness of the search result. This is to be distinguished from efficiency, which refers to the most economical use of system resources (runtime and memory efficiency). A standard approach in evaluating the effectiveness of an IR process is so-called benchmarking. According to the Duden definition, a benchmark is generally a "standard for comparing performance."

Historically, the approach is based on the so-called Cranfield experiments, which were conducted in the 1950s [9]. In these experiments, initial attempts were made to define the relevance of certain documents with respect to specific queries as a standard. However, due to disagreements among the developers, the project could not be fully completed. As part of the so-called Cranfield experiments by Cleverdon, formal tests of IR systems were actually conducted for the first time from 1958 to 1962 (Cranfield I). The aim was to index 18,000 documents from the field of aerodynamics using four different indexing methods and then evaluate how well information retrieval worked with queries in each case. Based on the insights gained, further experiments (Cranfield II) were conducted from 1962 to 1966. The approach practiced by Cleverdon has since been referred to as the Cranfield paradigm. The components of the experiment are commonly referred to as a test collection, which is described in Subsection 5.2.1. Section 5.2.2 then describes a procedure for its construction.

Benchmarking ensures comparability of different methods.

5.2.1 Test Collections

The experiments conducted as part of the Cranfield experiments were carried out in a controlled laboratory environment to ensure the comparability of multiple analysis results. According to the Cranfield paradigm, benchmarking consists of three essential components required for conducting an evaluation:

- **Fixed Set of Queries (Topics):** The queries are those submitted to the system during the tests. These are also referred to as topics, where a topic is intended to describe the user's information need and is usually more detailed. Query, on the other hand, refers to the translation of this information need into a query submitted to the IRS—these could be, for example, individual terms from the topic description. It is important that the topics are chosen to realistically reflect information needs.
- **Fixed Set of Documents (Corpus):** This refers to the underlying document base, also called the collection or corpus. These are the documents considered for a query—for example, scientific articles.
- **Fixed Set of Relevance Judgements:** This is the determination of the relevance of individual documents with respect to the various queries, which was initially purely binary, i.e., a document is either relevant or not relevant to a particular query. In the meantime, some assessments are also available that are not binary.

The classification of the documents in a test collection as relevant or not relevant is commonly referred to as the **gold standard** or ground truth. While the Cranfield collection was the first benchmark in the field of IR, various other benchmarks for testing IRS have emerged over time.

> Test collections increase reproducibility and evaluability.

5.2.2 Pooling

The queries of a test collection are created by experts from the domain to which the documents in the collection belong. In addition to providing queries, another requirement for the experts is to generate the gold standard. For this, a procedure called pooling is used. Pooling is introduced analogously to the description in [9], and the examples are also taken from this reference.

For relatively small collections, it is possible to manually assess all documents for their relevance. For large collections, this is no longer feasible. In this case, only a subset of the documents is assessed for relevance. This subset should not be created arbitrarily. Instead, a set of existing IRS is used, and their top-k results are merged and deduplicated. Using a single IRS would lead to an unbalanced document collection. This set of documents is then presented to the experts for assessment. The experts then evaluate which documents match the individual topics. Manning [2] points out in this context that this relevance assessment is made based on the topics, i.e., the information needs, and not the queries. A document that, for example, contains the terms from the queries is therefore not necessarily relevant. The experts' opinions are then consolidated according to a procedure to be defined, since it cannot be assumed that they will all make the same relevance judgments.

A gold standard can therefore also be evaluated by determining the agreement among the individual experts (inter-annotator agreement). For this, the so-called kappa statistic is used, which represents the agreement as a ratio:

$$kappa = \frac{P(A) - P(E)}{1 - P(E)} \tag{5.1}$$

$P(A)$ denotes the proportion of agreements and $P(E)$ the proportion of expected, random agreements. If more than two experts build the gold standard, an average pairwise agreement value is calculated. In general, a *kappa* value greater than 0.8 is considered good. A value between 0.67 and 0.8 is considered acceptable agreement, and a value below that is considered unusable.

> Pooling describes the construction of test collections.

5.3   Metrics for Evaluating Effectiveness

The preceding chapters described the construction and assessment of a test collection. This chapter addresses the question of how the effectiveness of an IR process can actually be calculated on this basis and compared with other evaluations.

In a benchmarking procedure, certain metrics are calculated based on the test collection. A selection of these metrics is introduced in the course of this chapter. First, a few basic facts are established. The documents in the collection can be divided, for each specific query, into relevant/not relevant and found/not found, as shown in Fig. 5.1, into different subsets. The desired outcome is for the IRS to produce only true positives and true negatives, of which, of course, only the true positives are output. This corresponds to a binary classification.

The following presents the most important IR metrics according to Croft [5].

5.3.1   Precision, Recall and P-R Diagrams

The two most important metrics for measuring effectiveness in IR are **precision** and **recall**. In general (and intuitively), recall indicates how well the IR process is able to find all relevant documents, while precision expresses how well the algorithm is able to exclude all non-relevant documents. To express this behavior as a numerical value, a set ratio of the total found and relevant documents is calculated as follows [10]:

$$Recall = \frac{G \cap R}{R} \tag{5.2}$$

$$Precision = \frac{G \cap R}{G} \tag{5.3}$$

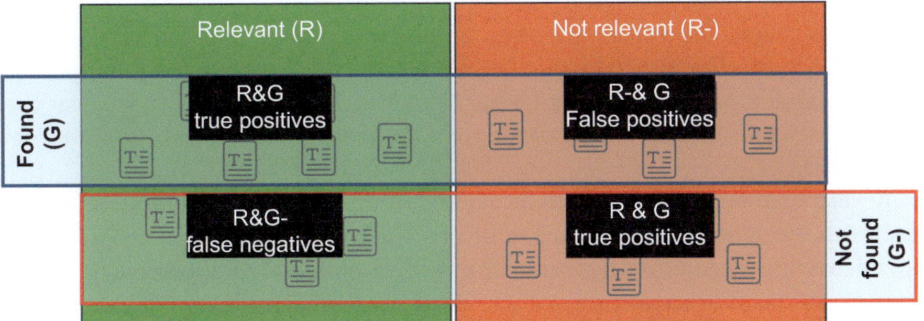

Fig. 5.1   Partitioning of the document set

Table 5.1 Precision and Recall—an Example

Rank	1	2	3	4	5	6	7	8	9	10
R/NR	+	−	+	−	−	+	−	−	+	+
Recall	1/5	1/5	2/5	2/5	2/5	3/5	3/5	3/5	4/5	5/5
Precision	1/1	½	2/3	2/4	2/5	3/6	3/7	3/8	4/9	5/10

Here, G denotes the set of retrieved objects and R the set of relevant objects. To obtain a complete assessment of the quality of a system or result, it is not sufficient to specify only precision or only recall, as each value alone is not meaningful. For example, to achieve a recall of 100%, one could simply return the entire document collection for every query—precision would, of course, be extremely low in this case. Conversely, to achieve high precision, only a few documents that are very likely to be relevant could be returned—in this case, recall would be very low. This makes it clear that precision and recall must always be considered together as a pair.

The calculation and visualization of precision and recall are explained using an example. Suppose there are five relevant documents in the entire document collection for a query, and the system returns the result set shown in Table 5.1, consisting of 10 documents ("+" stands for a relevant document and "−" for a non-relevant one).

Precision and recall can now be calculated for each rank within the result set. Since the values always indicate the state up to the respective document at the current position, precision is often referred to as *precision@K*, where K is the cut-off, i.e., the position up to which is considered—all subsequent documents are ignored. To make a statement about the complete result set, one would use the last value pair, in this example *Precision@10 = 0.5* and $R = 1$.

> Precision and recall are the most important standard metrics in IR.

For visualization, the values are often plotted in a precision-recall diagram (see Fig. 5.2). Connecting the individual points typically produces a sawtooth pattern, which is often removed by interpolation. Here, the precision at a drop is always set to the value that (looking further to the right in the diagram) was still achieved at a higher recall.

A very commonly used additional metric is the so-called **F-score**, which represents the weighted harmonic mean of precision and recall. If needed, it can be provided with an additional parameter α with $\alpha \geq 0$, which allows weighting in favor of precision or recall:

$$F = (a + \alpha) \cdot \frac{Precision \cdot Recall}{(\alpha \cdot Precision) + Recall} \tag{5.4}$$

For $0 \leq \alpha < 1$ this results in a weighting in favor of recall, while for $\alpha > 1$ the weighting shifts toward precision. In the case of $\alpha = 1$ an equal weighting for

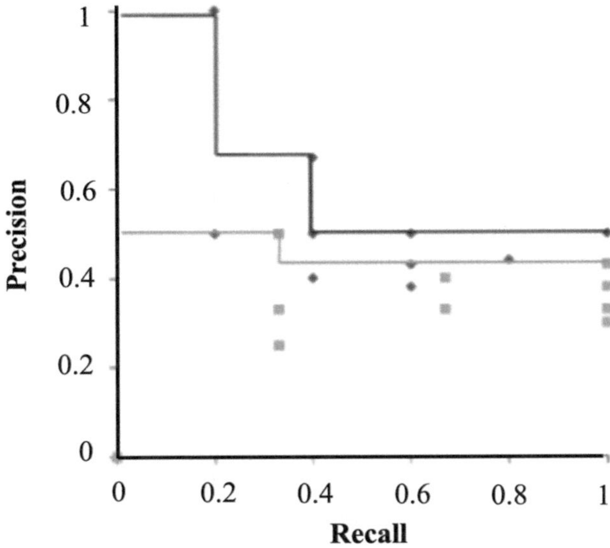

Fig. 5.2 Precision-recall diagram (from [6])

precision and recall is obtained. The F-measure is not a linear but a convex combination of precision and recall, i.e., as soon as one of the two values is very low, this will also lead to a low result.

5.3.2   Mean Average Precision (MAP)

While the previously introduced metrics only provided an evaluation based on a single query, mean average precision yields a value averaged over multiple queries. For a given query, $q_j \in Q$ applies, with $d_1, \ldots, d_m$ and R_{jk} being the set of all ranked retrieval results for Q.

For a single query ($|Q| = 1$), MAP calculates the average precision of all top-k documents (the set of all relevant documents up to position k). For more than one query, this value is normalized over the set of all queries:

$$MAP(Q) = \frac{1}{|Q|} \sum_{j=1}^{|Q|} \frac{1}{m} \sum_{k=1}^{m_j} Precision(R_{jk}) \qquad (5.5)$$

5.3.3   Reciprocal Rank and Mean Reciprocal Rank (MRR)

Reciprocal Rank (RR) is a metric that evaluates the first occurrence of a document. It corresponds to the reciprocal of the rank at which a relevant document first appears in the result list:

$$\frac{1}{K} \tag{5.6}$$

For example, if a result contains 100 documents and the first relevant document appears at rank 10, then the RR is $1/10 = 0.1$. Because of this property, this metric is often used to evaluate search methods that are expected to return only relatively few relevant results. Mean Reciprocal Rank (MRR) denotes the reciprocal rank averaged over a set of queries or the calculated RR values. For two queries q_1, q_2 with $RR(q_1) = 0.1$ and $RR(q_2) = 0.5$, the MRR is $(0.1 + 0.5)/2 = 0.3$.

5.3.4   Normalized Discounted Cumulated Gain (NDCG)

The metrics described so far referred to a binary result set, without allowing a ranking within the two classes. This is sufficient to evaluate all exact-match IR models. Best-match IR models, on the other hand, involve intrinsic uncertainty. The result of a search is output ranked by relevance, so that a ranking within the result set is also taken into account when determining a metric for evaluation. Discounted Cumulated Gain (DCG) is a measure that takes into account the ranking from highly relevant to less relevant or non-relevant documents in the evaluation. The calculation procedure is determined by the fact that up to a certain rank p, the top ranks of the output of an IR process accumulate positively and are reduced by lower ranks:

$$DCG_p = rel_1 + \sum_{i=2}^{p} \frac{rel_i}{\log_2 i} \tag{5.7}$$

The variable rel_i here denotes the RSV of the i-th rank. As an example, suppose the 10 RSVs of an output are given. The value of an RSV is distributed across four classes 0–3 (not relevant—highly relevant). The output of the retrieval process and the calculation of DGC_{10} are shown in Table 5.2.

However, evaluating DCG is not straightforward, as the measure is not normalized to a specific value range. This is remedied by normalization, which is calculated in the Normalized DCG (NDCG). The normalization factor corresponds to the ideal RSV (IDCG) for each rank (see also Table 5.3):

$$NDCG_p = \frac{DCG_p}{IDCG_p} \tag{5.8}$$

Table 5.2   Example Calculation DCG

Rank	1	2	3	4	5	6	7	8	9	10
RSV	3	2	3	0	0	1	2	2	3	0
Gain	3	2/1=2	3/1.59= 1.89	0	0	1/2.59= 0.39	2/2.81= 0.71	2/3= 0.67	3/3.17= 0.95	0
DCG@Rank	3	5	6.89	6.89	6.89	7.28	7.99	8.66	9.61	9.61

Table 5.3 Example Calculation NDCG

Rank	1	2	3	4	5	6	7	8	9	10
RSV ideal	3	3	3	2	2	2	1	0	0	0
IDCG	3	6	7.89	8.89	9.75	10.52	10.88	10.88	10.88	10.88
NDCG@Rank	1	0.83	0.87	0.76	0.71	0.69	0.73	0.8	0.88	0.88

NDCG is frequently used in scientific evaluations, as it provides a fairly good quantitative statement about the relevance of search results and the associated effectiveness of the IR process through a variety of upstream calculations.

> NDCG incorporates rank into the evaluation.

5.3.5   Evaluation Conferences

IR has long been an academic research field, with a number of established conferences where the latest research results are presented and discussed. Notable among these are the Text REtrieval Conference, CLEF (Cross Language Evaluation Forum), and NTCIR (NII Testbeds and Community for Information Access Research) [3]. Due to its pioneering role, this subsection will take a closer look at the TREC conference as an example.

TREC [3] is a "competition" that began in 1992 as part of TIPSTER (a project sponsored by the Defense Advanced Research Projects Agency, DARPA, to advance text processing methods for retrieval, summarization, and extraction), in which a large number of participants compete annually with their developed retrieval systems. TREC is sponsored, among others, by the National Institute of Standards and Technology (NIST) and the U.S. Department of Defense, and aims to support research in the field of IR. The TREC website specifically lists the following goals:

- Advancing research in IR based on large test collections,
- Improving communication between industry, academia, and government regarding the exchange of research ideas,
- Reducing time to market by highlighting improvements in retrieval methods with respect to their applicability to everyday problems,
- Improving the availability of suitable IR techniques as well as the development of new techniques.

TREC is overseen by a committee composed of representatives from industry, academia, and government. This committee also determines, based on the current research interests of the community, the so-called tracks that are offered each year. These are topic areas or research fields in which there is a need to conduct retrieval tasks. Over the years, various topics have been addressed through the offered tracks. In 2018, for example, there were the Common Core Track and the News Track, which dealt with searching news documents, and the Complex Answer Retrieval Track, which focused on answering more complex information needs. For the medical field, there has been the Clinical Decision Support Track (CDST) since 2014, which became the Precision Medicine Track (PMT) in 2017. This allows IR systems designed for medical professionals to be tested, providing relevant information regarding treated patients and their diseases. The current focus of the PMT is on cancer.

For each track offered, NIST compiles a corresponding test collection consisting of suitable topics and documents. As mentioned in the previous subsection regarding the Cranfield paradigm, the procedure is that all participants first apply their retrieval systems to the test collection and submit the highest-ranked results. In the so-called pooling process, the documents identified as relevant by the participants are pooled and assessed for their actual relevance. This produces the relevance judgments, which are then used to evaluate the individual systems. The test collections are also made available to non-participants for their own experiments and can be downloaded from the websites of the individual tracks. To conduct the evaluation of retrieval systems, TREC developed the program trec eval, which compares the results of a system under test with a gold standard and calculates various metrics. The program requires two input files: the query relevances, which represent the expected results, and the results generated by the system.

5.3.6 Datasets

The quality of the available input data plays an important role in the evaluation of IR processes. In the previous subsection, the TREC conference was already mentioned as an evaluation conference with a variety of pre-made datasets from different domains. However, TREC is a text-based evaluation conference, so the provided collections also consist of text documents. Especially in the field of general multimedia IR, additional collections are available that, for example, contain images, videos, composite social media posts, or audio files. Collections are often created for specific domains, such as the medical field (see Fig. 5.3).

Some datasets contain only multimedia objects that can be used for scientific purposes, while other datasets also include annotations that have been manually checked, normalized, and aligned with existing MMIR systems, thus serving as a gold standard. Examples of such datasets include:

- Flickr30k: a collection of smaller images with manually annotated metadata [11]
- DIV2K: a collection of high-resolution images without metadata [12]
- PASCAL VOC: a compilation of images based on their contained objects [13]
- HILOD21: a high-resolution and fully annotated image dataset created as part of research at FernUniversität in Hagen [14]

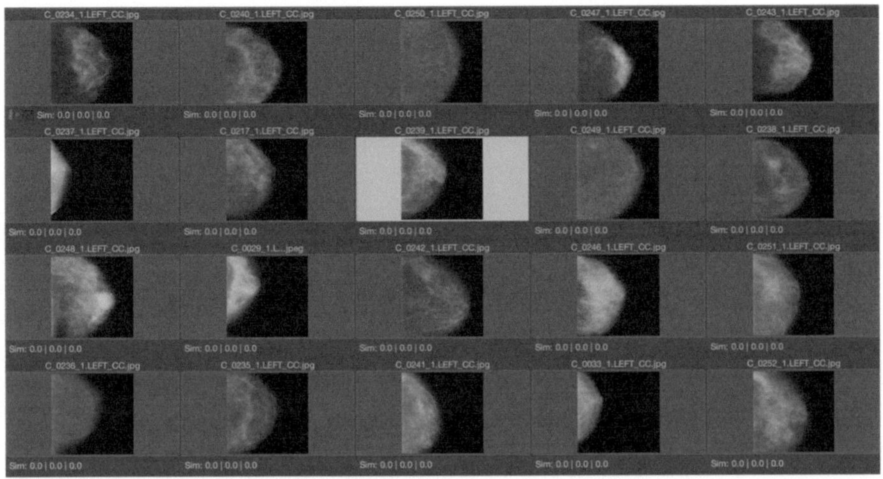

Fig. 5.3   Medical collections [1]

- FMA: a free music archive with metadata [15]
- Million Song Dataset: as the name suggests, a collection of 280 GB of audio files [16]
- LibriSpeech: a collection with approx. 1000 hours of English speech [17]
- VoxCeleb: a collection for speaker identification [18]
- Mivia Audio Events: 6000 different sounds from surveillance applications [19]

This small selection shows that there are datasets with varying characteristics and quality for many different domains. As an example, the HILOD21 dataset is briefly used here to illustrate which predefined annotations of images can be expected and against which information one can check their own evaluations.

Figure 5.4 shows some examples from the HILOD21 dataset. This dataset stands out because a large number of high-resolution images were selected and annotated both manually and automatically. The machine annotation was then manually checked. As a result, the images in this dataset have very accurate and valid metadata. An example of this is shown in Fig. 5.5.

5.3.7   Benchmarking

A comprehensive critical examination of benchmarking in IR is discussed by Voorhees in [20]. This chapter addresses some of the criticisms raised by Voorhees. The application of benchmark evaluation represents a generalization of the actual retrieval being assessed. The underlying assumption is that a successful computation of a ranking corresponds to a good evaluation of the tested IR process. However, the users of IRS are the actual target group who should be evaluating the results. After

Fig. 5.4 Examples from the HILOD21 dataset [14]

all, they are the true beneficiaries of automated retrieval. Evaluations with users, however, are expensive and labor-intensive. Some reasons for this include:

- A large number of participants is required, who must be found and whose interest/willingness must be engaged.
- If IRS are compared based on the benchmark, then the interfaces for communicating with the system must also be comparable, so that results are not distorted at this level.
- Participants must be equally trained in the use of IRS.

The reasons mentioned above, in particular, argue for the use of a system-oriented evaluation such as benchmarking, which traces back to the Cranfield experiments. The Cranfield experiments are based on three assumptions made for simplification:

Attribute	Value
Name	ab307b8f-662b-4429-9123-804a73e1038a.png
Keywords	library, building, congress of the united states
Primary Objects	library
Other Objects	building, window
Description	library of congress of the united states of america. The image is taken inside the building and shows seats for reading in a large hall.
Automated detection	fixture, interior design, wall, symmetry, facade, real estate, glass, landmark, city

Fig. 5.5 Metadata of the HILOD21 dataset [14]

- The calculation of thematic proximity determines the relevance of a document to the query. This assumption implies that every relevant document is equally desirable in the output of an IRS, the relevance of a document is independent of the relevance of all other documents, and that the information need is static in nature.
- The gold standard is representative of all users of an IRS.
- The gold standard is complete with respect to all potentially relevant documents.

These strong simplifications also make benchmarking a noisy process. The assumption that every relevant document has the same relevance seems to be a strong simplification. However, a very intuitive assumption is that different documents on a topic can be very different—in their structure, depth, or the perspective from which a topic is addressed. The assumption that the information need is static also seems artificial, as it does not take into account the dynamic processes of system users, who gain knowledge while viewing the result list and, based on the first documents viewed, may already have a changed assessment regarding the relevance of documents. By successively viewing documents classified as relevant, the perspective on the relevance of the available documents changes, and the information need cannot be considered static.

Common test collections, in terms of the number of documents they contain, exceed what is feasible when it comes to manually reviewing a collection for all relevant documents and including them in the gold standard. So-called pooling was introduced as a solution in Sect. 5.2.2. However, the gold standard is effectively based only on a subset of the entire document collection. There is concern that not all relevant documents have been included in the gold standard. Unjudged documents can strongly influence the calculation of metrics.

Another criticism of the creation of the gold standard is inconsistency. This inconsistency is caused by the individuality of experts' assessments of whether a document is relevant or not. Chapter 5 has highlighted some facets of this concept and shown how multifaceted the concept of relevance is. One way to address this challenge is to obtain multiple expert opinions on the relevance of documents and to have all documents reviewed by several experts in a multi-stage process. Section 5.3.2 described how the kappa statistic can be calculated as an indicator of the level of agreement among contributing experts on a topic. Based on this calculation, it must be decided whether the gold standard can be meaningfully used in an evaluation.

5.3.8   Application and Interpretation of Metrics

Although the metrics presented here are widely used in the state of the art and practice to substantiate effectiveness analyses, there are also some criticisms that should be discussed at this point. The following discussion is based on Fuhr's publication [21] and is initially limited to the use of MRR.

MRR calculates the average reciprocal value of the first occurrence of a relevant document. However, the significance of this result should be critically examined. Fuhr points out an unintended behavior of this metric in [4]. Suppose an IRS A returns the first relevant hits for queries at ranks 1, 2, and 4, and an IRS B returns the first relevant hits each at rank 2. In this case, the average first hit of IRS A is at rank $\frac{1+2+4}{3} = 2.334$. For IRS B: $\frac{2+2+2}{3} = 2$. Thus, the average value for IRS A is worse than that for IRS B.

How do the two result lists behave when evaluated using MRR? Here, for $MRR(A) = \frac{1+\frac{1}{2}+\frac{1}{4}}{3} = 0.58$, but for $MRR(B) = \frac{\frac{1}{2}+\frac{1}{2}+\frac{1}{2}}{3} = 0.5$. Although, on average, IRS B places its results at lower ranks, the calculation of MRR indicates the opposite. With respect to MRR, IRS A is thus more effective than IRS B.

5.4   Summary

In this chapter, we have examined in detail the possibilities for assessing the quality/effectiveness of MMIR methods. To introduce reproducible and measurable procedures, the topic of test collections was first described, which remains an important method for creating representative reference data and datasets. On this basis, metrics can then be introduced, of which precision and recall are certainly the most important and widely used. These metrics help to evaluate the quality of the implemented MMIR methods. It is interesting to note that a number of standardized procedures and measurements have been established in both research and practice, such as NDCG. On this basis, the effectiveness of MMIR methods can now be measured, compared with other methods, and, in the best case, optimized.

5.5   Self-Assessment Questions

4.1—Test Collections

What criteria must be met for test collections according to the Cranfield paradigm?

4.2—Metrics

How are precision and recall defined?

4.3—Metrics

What is NDCG used for?

References

1. I. Maglogiannis, C. Doukas, G. Kormentzas, and T. Pliakas, "Wavelet-based compression with roi coding support for mobile access to dicom images over heterogeneous radio networks," *Information Technology in Biomedicine, IEEE Transactions on*, vol. 13, pp. 458–466, 2009. DOI: https://doi.org/10.1109/TITB.2008.903527.
2. C. D. Manning, P. Raghavan, and H. Schütze, *Introduction to Information Retrieval*. Cambridge University Press, 2008.
3. The Text Retrieval Conference (TREC). "Datasets." [Online]. Available: https://trec.nist.gov/data.html, Download: 11.10.2021.
4. R. Baeza-Yates, B. Ribeiro-Neto, and N. Fuhr, *Information Retrieval: Implementing and Evaluating Search Engines*. MIT Press, 2011.
5. F. Song and W. B. Croft, "A general language model for information retrieval," in *Proceedings of the Eighth International Conference on Information and Knowledge Management*, ser. CIKM'99, Kansas City, Missouri, USA: Association for Computing

Machinery, 1999, 316–321, ISBN: 1581131461. DOI: https://doi.org/10.1145/319950.320022. [Online]. Available: https://doi.org/10.1145/319950.320022.

6. W. B. Croft, D. Metzler, and T. Strohman, "Search engines – information retrieval in practice," 2009. [Online]. Available: https://api.semanticscholar.org/CorpusID:2350758.

7. Wilson and Sperber, "Relevanz: Eine interdisziplinäre geschichte," Jan. 2006, Fachhochschule and Nijmegen. DOI: https://doi.org/10.13140/2.1.1300.8645.

8. T. Saracevic, "Relevance reconsidered," in *Information science: Integration in perspectives. Proceedings of the Second Conference on Conceptions of Library and Information Science, Copenhagen (Denmark)*, 1996.

9. E. M. Voorhees, "The evolution of cranfield," in *Information Retrieval Evaluation in a Changing World: Lessons Learned from 20 Years of CLEF*, N. Ferro and C. Peters, Eds. Cham: Springer International Publishing, 2019, pp. 45–69, ISBN: 978-3-030-22948-1. DOI: https://doi.org/10.1007/978-3-030-22948-1_2. [Online]. Available: https://doi.org/10.1007/978-3-030-22948-1_2.

10. D. AI. "Precision and recall," Deep AI. [Online]. Available: https://deepai.org/machine-learning-glossary-and-terms/precision-and-recall, Download: 11.08.2021.

11. P. Young. "From image descriptions to visual denotations: New similarity metrics for semantic inference over event descriptions." [Online]. Available: http://shannon.cs.illinois.edu/DenotationGraph/, Download: 01.10.2020.

12. E. Agustsson and R. Timofte, "Ntire 2017 challenge on single image super-resolution: Dataset and study," in *The IEEE Conference on Computer Vision and Pattern Recognition (CVPR) Workshops*, 2017. [Online]. Available: http://www.vision.ee.ethz.ch/~timofter/publications/Agustsson-CVPRW-2017.pdf.

13. M. Everingham, L. Van Gool, C. K. I. Williams, J. Winn, and A. Zisserman, "The pascal visual object classes (voc) challenge," *International Journal of Computer Vision*, vol. 88(2), pp. 303–338, 2010.

14. S. Wagenpfeil. "A high level-of-detail image dataset." (Nov. 2020) [Online]. Available: http://www.stefan-wagenpfeil.de/public/GMAF/.

15. M. Defferrard, K. Benzi, P. Vandergheynst, and X. Bresson, "Fma: A dataset for music analysis," in *Proceedings of the 18th International Society for Music Information Retrieval Conference (ISMIR)*, 2017. [Online]. Available: https://arxiv.org/abs/1612.01840.

16. T. Bertin-Mahieux, D. P. W. Ellis, B. Whitman, and P. Lamere, "The million song dataset," in *Proceedings of the 12th International Society for Music Information Retrieval Conference (ISMIR)*, University of Miami, 2011, pp. 591–596. [Online]. Available: http://ismir2011.ismir.net/papers/OS6-1.pdf.

17. V. Panayotov, G. Chen, D. Povey, and S. Khudanpur, "Librispeech: An asr corpus based on public domain audio books," in *Proceedings of the IEEE International Conference on Acoustics, Speech and Signal Processing (ICASSP)*, IEEE, 2015, pp. 5206–5210. DOI: https://doi.org/10.1109/ICASSP.2015.7178964. [Online]. Available: http://www.openslr.org/12.

18. A. Nagrani, J. S. Chung, and A. Zisserman, "Voxceleb: A large-scale speaker identification dataset," *arXiv preprint* arXiv:1706.08612, 2017. [Online]. Available: https://arxiv.org/abs/1706.08612.

19. P. Foggia, N. Petkov, A. Saggese, N. Strisciuglio, and M. Vento, "Reliable detection of audio events in highly noisy environments," *Pattern Recognition Letters*, vol. 65, pp. 22–28, 2015. DOI: https://doi.org/10.1016/j.patrec.2015.06.026.

20. E. M. Voorhees, "Coopetition in IR research," in *Proceedings of the 43rd International ACM SIGIR conference on research and development in Information Retrieval, SIGIR 2020, Virtual Event, China, July 25-30, 2020*, J. X. Huang et al., Eds., ACM, 2020, p. 3. DOI: https://doi.org/10.1145/3397271.3402427. [Online]. Available: https://doi.org/10.1145/3397271.3402427.

21. N. Fuhr, "Probabilistic models in information retrieval," *The computer journal*, vol. 35, no. 3, pp. 243–255, 1992, Publisher: The British Computer Society.

22. Cambridge Dictionary. (n. d.). *Relevance*. In English–German Dictionary. Cambridge University Press. Abgerufen am 3. April 2024 von https://dictionary.cambridge.org/dictionary/english-german/relevance
23. Dudenredaktion. (n. d.). *Relevanz1*. In Duden Online. Bibliographisches Institut GmbH. Abgerufen am 3. April 2024 von http://www.duden.de/rechtschreibung/Relevanz
24. Merriam-Webster. (n. d.). *Relevance*. In Merriam-Webster.com Dictionary. Abgerufen am 3. April 2024 von https://www.merriam-webster.com/dictionary/relevance

Applications and Outlook

At the beginning of the book, we stated "Multimedia is everywhere," and we would like to revisit this statement here in this final chapter. Now that you have engaged extensively and intensively with the concepts of Multimedia Information Retrieval and have gained a deeper understanding of the underlying models, algorithms, concepts, and technologies, we now place the question "what is all this for?" at the center and will first present a series of applications from the field of MMIR, some of which might not initially be recognized as such. In addition, we will contrast MMIR with applications from the field of generative AI and conclude with an outlook on virtual and augmented worlds.

6.1 Multimedia Information Retrieval Applications

As mentioned in the introductory subsections, nearly every application can be traced back to MMIR processes. Clicking on a link, for example, is a way to satisfy the information need of users (who ultimately want to see the next page). Therefore, despite the term "Multimedia IR," one should not assume that such applications always appear particularly multimedia-heavy. Of course, classic search engines, information systems, databases, etc. play an important role, and naturally, applications such as video streaming or desktop search are classic MMIR applications. Apps from the fields of social media, collaboration, and interaction can also be mapped to MMIR processes, but there are also a number of applications that can only be identified as MMIR applications at second glance. In this subsection, we will present some of these rather exotic applications as examples, in order to showcase the broad application area of MMIR and the wide range of uses for the concepts presented here.

© The Author(s), under exclusive license to Springer-Verlag GmbH, DE, part of
Springer Nature 2026
M. Hemmje and S. Wagenpfeil, *Multimedia Information Retrieval*,
https://doi.org/10.1007/978-3-662-73310-3_6

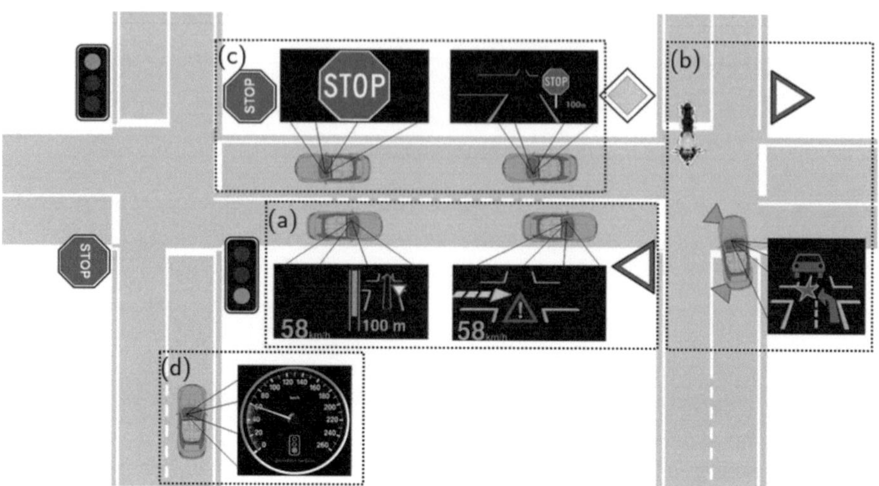

Fig. 6.1 Different representations of intersection assistants in the automotive sector

6.1.1   Automotive Applications

In the automotive sector, for example, cockpits are becoming increasingly digital. Most modern vehicles feature fully digital representations of the current vehicle status. Even the display of the current speed satisfies the information need of a user—although it can be debated whether this is always an active, user-initiated process. In any case, so-called assistance systems can be considered MMIR components, as they support users in perceiving information about their surroundings and evaluating their features. A good example of this is so-called intersection assistants, which display a range of information to the driver when entering an intersection (see Fig. 6.1).

In this case, the information need is implicitly assumed by the vehicle, as the user is approaching a safety-relevant area. Feature extraction then takes place via image recognition, object detection, radar, or even audio sensors (e.g., an approaching siren). The features thus detected are processed in an internal (or sometimes even cloud-based) MMIR system and analyzed using feature fusion or weighting. As soon as a certain result is detected (e.g., a vehicle approaching from the left), the MMIR system generates a corresponding response, which is then visualized in the cockpit. The greatest challenge in this area is real-time processing and thus the most efficient possible representation and processing of MMIR features. The aspect of archiving, however, can be neglected here.

6.1.2   Applications on Smart Devices

Smart devices, such as smartwatches, record a multitude of sensors. Step counters, heart rate measurement, blood oxygen levels, and much more have now become standard for such devices. All these sensor data can be interpreted as MMIR features. If they are recorded for further processes, this reflects the typical ingest process in the IR model. When the smartwatch data is transmitted to the associated smartphone, these are stored there (or in the cloud) as MMIR features. The subsequent viewing of the corresponding information follows the satisfaction of the information need—usually in the form of dashboards and overview graphics (see Fig. 6.2).

Such data can now also be shared with treating physicians—which can be mapped by the distribution of MMIR processes. From the physician's perspective, an information need is again represented, and here too the MMIR processes start anew.

6.1.3   Gaming Applications

Computer games can be viewed as large, distributed MMIR applications. Every user who acts in the game generates MMIR features through their actions, their

Fig. 6.2 Example display of fitness data recorded with a smartwatch

interactions with the game, or with other users. The challenge for the game back-end (in this interpretation, the central MMIR system) is to manage the multitude of features from all players and their interactions and to derive the respective information need of each player. This is then presented in the form of the next scene or subsequent interaction—a classic MMIR process.

6.1.4   Applications on Self-Service Terminals

Recently, more and more self-service terminals are being installed in retail stores. These too are classic MMIR applications. Using a smartphone app or QR code scanner, for example, the products a customer places in their shopping cart are recorded. This is again a classic ingest process (as is the case with any shopping application). The representation of products here is even simpler than in the automotive sector, because they do not need to be recognized, but their item number (i.e., the digital feature representation) can be determined directly via barcode or QR code. On the way to the exit, the self-service terminal is located, where the payment and checkout process is started. Here, the user's information need is again interpreted, in this case the desire to pay for the products in their shopping cart. The MMIR system performs a search for relevant information (prices, total calculation, etc.) and presents the result to the user (in the simplest case, the amount to be paid). The subsequent payment process can also be mapped analogously via MMIR processes (Fig. 6.3).

These selected examples show that MMIR is used in many areas—even in those that may initially seem unrelated to the field. More and more, applications are being supplemented by the use of digital assistant functions, which currently culminate in the field of generative AI. The following subsection provides further information on this and a distinction from MMIR.

Fig. 6.3 Self-service terminals in retail [6]

Multimedia Information Retrieval is everywhere!

6.2   Generative AI

In Subsection 2.6, some concepts of Artificial Intelligence and Machine Learning have already been introduced. An increasingly important aspect here is so-called Generative AI (GenAI), which makes it possible to create a wide variety of new content based on existing information and specifications. This is a fundamentally different approach from MMIR, as the content "sought" by users is not found, but newly generated. For example, if users want to find an image with the textual query "beautiful venice canals with gondolas and bridges," classic MMIR methods would search the collection for suitable multimedia objects and return a list of results. Generative AI, on the other hand, computes an artificial result, which is then presented to the user. If you enter the query mentioned here (in this context also called a prompt) into various generative AI systems for image generation, surprisingly good results are produced. Figures 6.4, 6.5 and 6.6 illustrate this impressively. It should be noted again that none of these images depicts a real existing scene, but rather they are artificial computations based on trillions of training data.

Fig. 6.4 Response generated with Midjourney [3]

Fig. 6.5 Response generated with Stable Diffusion [7]

Fig. 6.6 Response generated with Dall-E [5]

Similarly, AI-generated texts are not real documents (or information) that are returned, but inventions based on training data, which in many cases are surprisingly good and accurate, but can also be arbitrarily incorrect. Nevertheless, these first examples already show that for users of multimedia applications, the boundary between MMIR and generative AI quickly becomes blurred. Users enter a query and receive a result. The quality of the result is so good that users cannot distinguish it from genuine results. Due to years of experience with classic MMIR systems (such as search engines), users also initially assume the correctness of the presented information and do not question its origin. To make matters worse, generative AI is not explainable, i.e., even the manufacturers of the systems can no longer trace which training data or connections led to the corresponding result. The following subsections briefly introduce some systems and considerations on this topic.

6.2.1 Generative AI Systems

The release of Chat-GPT [1] in 2022 triggered a veritable boom in generative AI systems. According to current estimates, the impact on the global economy could create added value of $2.6 to $4.4 trillion per year by increasing productivity in many use cases. Thus, generative AI systems are not just a technical gimmick, but one of the most relevant technologies of our time. The development, integration, and application of generative AI methods therefore also represent a core area of computer science, information systems, and many related disciplines. Due to the multitude of systems and the rapid release of new use cases, a complete listing is impossible. Nearly all major manufacturers, such as Microsoft, Facebook, Google, Amazon, Apple, Adobe, have more or less mature and powerful generative AI systems. Currently, however, there are a number of systems that can serve as selected examples:

- **DALL-E** 2 [4] is an AI developed by OpenAI that can create realistic images and art using text to image.
- **Stable Diffusion** [7] is an AI developed by StabilityAI that can also generate images based on natural language expressions. Compared to DALL-E 2, Stable Diffusion is open-source software.
- **Midjourney** [3] is an AI that can create images and art based on a text description.
- **GPT-3.5** [1] and **GPT-4** are language models capable of performing a range of tasks based on instructions in the form of natural language expressions.
- **Google Vertex AI** is a machine learning platform. The Vertex AI API provides access to Google's PaLM 2 [8], a family of language models. Specializations of PaLM 2 include Med-PaLM for medical use cases and Sec-PaLM 2 for security-related use cases. PaLM 2 powers generative AI features such as the PaLM API, tools, and Bard.

The representatives presented here focus on images and text. But in the category of audio generation programs, there are also well-known systems such as those from Eleven Labs, Resemble AI, or Bark. Use cases for these systems can include multilingual dubbing, voice cloning, or automatic generation of subtitles in videos. In the category of video generation programs, well-known systems include those from RunwayML, Synthesia, or Rephrase AI. Use cases for these types of systems can be rapid video generation, video editing, especially in marketing or news broadcasting. Audio or video systems are neglected in the following due to the limited possibilities for illustration in a book. However, the same rules and conditions apply to them as to images and texts.

6.2.2   Framework Conditions and Limitations

Current generative AI systems are capable of producing convincing and realistic-looking content. Nevertheless, these systems are subject to a number of limitations:

- **Transparency and Explainability:** Generative AI uses trillions of training data, from which models are built via machine learning and then used to generate results. The entire process of learning and thus the shaping of the model must be considered a black box due to the amount of data that is no longer comprehensible to humans. Insight into the decision-making and generation processes that take place within the AI is not possible for humans. It is not apparent how a model arrives at a decision and a result. This means that the results and their production are not explainable, and the entire process is therefore not transparent. This not only affects the understanding of generative AI processes but also their operational use in various application areas. Systems that produce information using generative AI, for example, cannot be certified because the algorithms and rules that lead to the production of the information cannot be documented.
- **Dependence on Training Data:** Results generated with generative AI are the result of their training data. The selection of training data therefore largely determines the later characteristics of the models and the results produced. Due to the lack of transparency, inaccuracies in the training data can also have major effects on the later results. Therefore, the models are usually adapted to specific use cases and, as a consequence, are also difficult to transfer to other areas.
- **Ethics:** If generative AI is supplied with unbalanced training data, certain "opinions" can be reinforced. In extreme cases, the models can even generate harmful content. Especially in view of various ethical debates of our time, providing balanced training datasets is a highly sensitive issue. Just as in the real world, every group wants to be represented in the training data so that the later results also contain the respective opinions of the group. From an ethical perspective, this seems to be an almost unsolvable problem.

- **Data Protection:** In the ongoing discussions on the topic, there are repeated cases of sometimes massive data protection violations. In some cases, the generated results can indirectly reveal information about the training data, for example, because a specific term is queried. This term must, of course, have appeared in the training data if the resulting model provided a plausible answer. It has now been found that in some cases, documents have been included in the training data for which the authors did not explicitly grant permission for use. Since much training data is collected via web crawlers or social media APIs, data protection violations in connection with generative AI are quite likely.
- **Legal Framework Conditions:** Two major topics play a role here. On the one hand, the question of the use of input data, which was already explained in the previous point. On the other hand, the question of rights to the generated result. Who "owns" the image in Fig. 6.4, for example? Does it belong to the company whose model generated the image? Or does it belong to those who provided the associated training data? Where may this image be used? What if the image depicts people or actions that never actually took place (keyword "fake news")? What if this causes damage—who is liable? How can (and possibly must) works generated by AI be labeled? Did the student really write the paper themselves or have it written by generative AI? All these are questions that are currently still unresolved and for which reliable and implementable solutions must be found.
- **Context Sensitivity:** Generative AI also requires appropriate context to resolve the term "Jaguar." This is typically provided through descriptions. The more precise the descriptions, the better a desired result can be generated. The context can become arbitrarily complex and extensive—in some use cases, it must be. In practice, this can lead to very elaborate textual descriptions of the query (so-called prompts).

> There are numerous unresolved issues in the field of generative AI.

Generative AI will find its way into many areas of daily life and the professional world. Most IT systems already have intelligent assistants that generate texts as needed, modify images, perform translations, or find errors in documents (even source code). A good example of this is Adobe Firefly [2], which extends the image editing software Photoshop with generative AI. Figure 6.7 shows this. Using the textual input "yellow road lines," generative AI can be applied to the selected area of the image, which then creates a new layer in the image with the corresponding additions.

At the latest here, the question arises again as to the author and rights holder of this generated image. Fortunately, this is not a topic we need to address within the scope of this book. However, it remains exciting to see how such issues will be resolved in the future.

Fig. 6.7   Generative AI in Adobe Photoshop [9]

6.3   Virtual and Augmented Reality, Metaverses

Although the field of Virtual/Mixed/Augmented Reality (VR/MR/AR) and the so-called metaverses may seem like a recent development, as early as 1986 the company Lucasfilm released one of the first virtual worlds called Habitat [10]. At that time, personal computers such as the Commodore 64 were popular, and the infrastructures for data exchange were just beginning to emerge.

Habitat was an example of a so-called Massively Multiplayer Online Game (MMOG), in which users could create avatars and interact in a digital world. Due to technical limitations at the time, not enough users were able to participate in such games, which led to the project being discontinued in 1988. However, even in this first example of a virtual world (see Fig. 6.8a), it is clear that this is a multimedia application with images, text, sound, and animations, and that dynamic content is naturally also presented there, which must somehow be determined.

Fig. 6.8 Screenshots of various virtual worlds. **a**) Habitat 1986 [10], **b**) the Metaverse by Meta [11], **c**) a Microsoft Teams conference with avatars [12], **d**) the MILC Metaverse [13]

Thus, Habitat already utilized the fundamental processes and concepts of MMIR systems.

Many of the concepts anticipated back then are now implemented in computer games and virtual worlds (also called metaverse, see Fig. 6.8b–d). In particular, Fig. 6.8c is an exciting example of the integration of real and virtual content, as it shows a Microsoft Teams chat in which individual users are represented only by their avatars, which imitate the gestures and facial expressions of the respective users. Enormous advances in computer graphics and the availability of ever faster graphics processors for rendering 3D worlds and capturing real environments have led to a variety of use cases.

The field of **Augmented Reality** refers to the projection of enhancements onto real worlds. Figure 6.9 shows an example in which the user employs a tablet, where the current live image from the built-in camera is enriched with additional elements on the display.

In the field of **Mixed Reality**, it is assumed that the user perceives the 3D world not just through a small device such as a tablet or smartphone, but through a suitable 3D headset that provides a see-through view of the real world, onto which virtual elements are digitally superimposed (see Fig. 6.10).

Complete virtual worlds project a fully digital image of any real or virtual location using real-time 3D visualization, either onto a screen or—typically—directly onto the user's retina via a 3D headset. In such environments, any real or virtual representations can then be consumed (see Fig. 6.11).

With these technologies and the bandwidths currently available, an incredible variety of applications is, of course, possible. Without going into technical

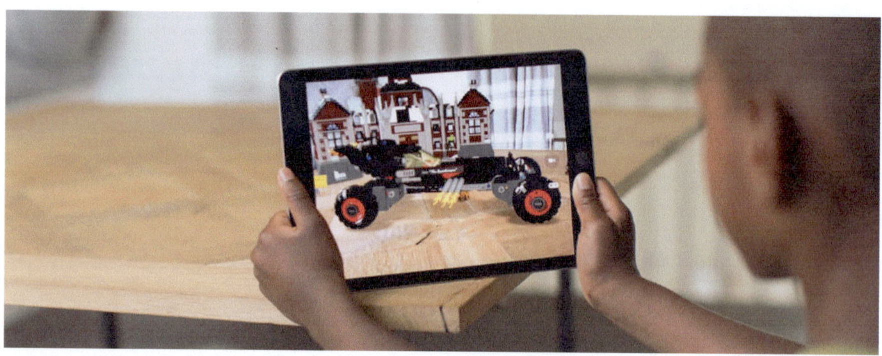

Fig. 6.9 An augmented reality application on the Apple iPad [14]

Fig. 6.10 A mixed reality application from Microsoft [12]

details here, it is nevertheless necessary at this point to establish the connection to Multimedia Information Retrieval.

As has already been mentioned several times, virtually every type of application and interaction can be described and mapped using MMIR processes. Clicking a "Like" button is just as much a part of this as entering a virtual portal. For example, when a user enters a room in a virtual world, this can be interpreted as an information need. The user is specifically searching for information or browsing

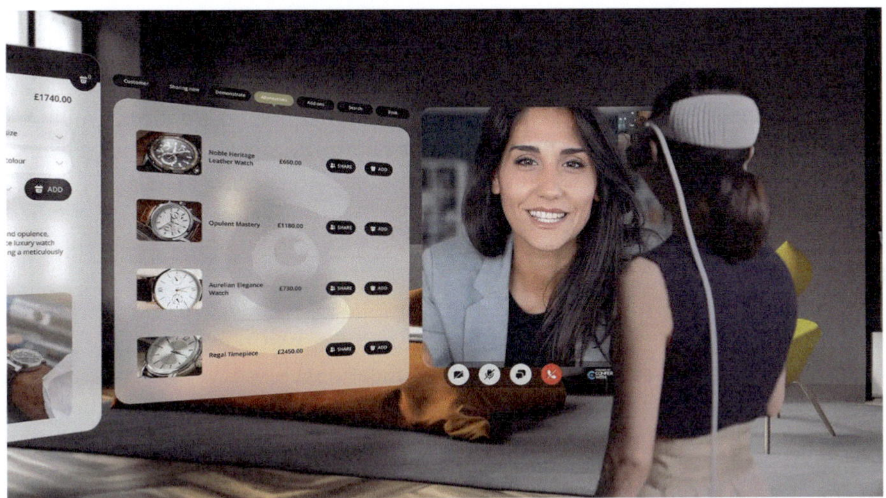

Fig. 6.11   A virtual reality application from Apple on the Vision Pro [15]

around the room (Searching vs. Browsing). The system that generates the virtual or augmented world is essentially a large MMIR system whose task is to satisfy this information need. Even reaching the next level in a computer game can be viewed in this way as an information need; the user's efforts to reach it then represent the complexity of formulating the information need, and jumping to the next level is the satisfaction of that need. Objects displayed in the virtual or augmented world are the results of MMIR processes, because when a user, for example, wants to recall a virtual experience, nothing other than an MMIR search is performed in the background. There are two key differences at this point:

- **Query construction:** Queries to virtual or augmented worlds are typically made using additional modalities. For example, gestures are introduced to replace mouse clicks or finger taps on smartphones. Voice input is also increasingly used. Since one can also move away from the classic UI concept of windows or apps, any objects can now be interactive and thus represent the information need and also trigger MMIR processes.
- **Result presentation:** The diverse possibilities in digitally augmented or rendered worlds also have an impact. Results can be animated, displayed, or made interactive in any way. Typically, a mix of the multimedia types presented in this book is used for this purpose.

VR/AR/MR applications are also MMIR applications.

Most other processes in these worlds can also be directly represented by MMIR processes. This is particularly interesting because it means that the complexity of VR/AR applications and metaverses can essentially be reduced to the differences mentioned above. Once again, this area of application demonstrates that multimedia is everywhere and that Multimedia Information Retrieval forms the universal basis for virtually every application.

6.4 Summary

This final chapter rounds off the field of MMIR and first presents a series of use cases that go beyond the "typical" MMIR applications. Since MMIR does not only take place on computers, but in virtually every digital presentation of media—everywhere a display is hidden—there are countless other examples not listed in this section. We would like to encourage you at this point to find these and to discover MMIR applications with open eyes and ears. However, it is important to make a distinction with regard to generative AI, which was discussed in Subsection 6.2. While MMIR systems retrieve results from a collection—thus returning existing knowledge—generative AI systems create new knowledge, regardless of whether it is correct. For users accustomed to "trusting" MMIR systems, this poses a significant risk, as photos generated by Stable Diffusion, Midjourney, or Dall-E no longer show real scenes; what you see in them does not exist. And unfortunately, such photos can no longer be distinguished from real photos. This issue can be further amplified by virtual and augmented applications, in which the boundary between reality and fiction is deliberately crossed. The integration of MMIR with generative AI and VR/AR is an exciting and highly topical area of research.

References

1. Open.AI, *Chat gpt*, https://chat.openai.com/, 2023.
2. Adobe Systems Software Ireland Ltd. "Creative journeys start here." [Online]. Available: http://www.adobe.com.
3. I. Art, *Midjourney*, https://www.imagine.art/, 2023.
4. Open.AI, *Dall-e*, https://openai.com/dall-e-2, 2023.
5. OpenAI, *DALL.E: Creating Images from Text*, https://openai.com/dall-e, Zugriff; am 9. Juni 2025, 2025. [Online]. Available: https://openai.com/dall-e.
6. Trendwelten Redaktion. "Self-Checkout: Kontaktlos mobil bezahlen." Zugriff; am 9. Juni 2025. [Online]. Available: https://www.trendwelten.eu/trend/self-checkout-kontaktlos-mobil-bezahlen/.
7. Stablity.AI, *Stable diffusion*, https://stabiliti.ai, 2023.
8. Google.com. "Google Vision AI – derive insights from images." [Online]. Available: http://cloud.google.com/vision, Download: 04.11.2021.
9. Adobe Inc., *Adobe Photoshop*, https://www.adobe.com/products/photoshop.html, Zugriff; am 9. Juni 2025, 2025. [Online]. Available: https://www.adobe.com/products/photoshop.html.
10. C. Morningstar and F. R. Farmer, *Habitat*, Lucasfilm Games / Quantum Link, Beta-Version für den Commodore 64, 1986. [Online]. Available: https://archive.org/details/LucasfilmGamesHabitat.

11. Meta Platforms, Inc., *Introducing meta: A social technology company*, https://about.fb.com/news/2021/10/facebook-company-is-now-meta/, Pressemitteilung zur Umbenennung von Facebook Inc. in Meta Platforms Inc. und zur strategischen Ausrichtung auf das Metaverse, 2021.
12. Microsoft. "Microsoft inc.," Microsoft Inc. [Online]. Available: https://www.microsoft.com, Download: 11.07.2021.
13. MILC Platform, *Milc metaverse: The open web3 media & entertainment platform*, https://www.milc.global/en, Entwickelt von Welt der Wunder TV, unterstützt durch Unreal Engine und Web3-Technologien, 2024.
14. Wikipedia. "Apple m1." [Online]. Available: https://en.wikipedia.org/wiki/Apple_M1, Download: 11.08.2021.
15. Apple Inc., *Introducing apple vision pro: Apple's first spatial computer*, https://www.apple.com/newsroom/2023/06/introducing-apple-vision-pro/, Pressemitteilung zur Vorstellung des Apple Vision Pro auf der WWDC 2023, 2023.

Appendix

Solutions

A.1 Solutions to Chap. 1

1.1—Information

According to Kuhlen, what is the relationship between information, knowledge, and action?

According to Kuhlen, information is knowledge in action, i.e., through a specific action or the information need of a user, relevant information is generated from knowledge.

1.2—Information Need

What is the pattern called when the information need is described by providing an example?

Query by Example

1.3—SECI Model

What phases does the SECI model have and what do they represent?

- Socialization: Transfer and processing of knowledge through explanation or demonstration
- Externalization: Documentation of knowledge, conceptualization
- Combination: Systematic analysis of knowledge
- Internalization: Processing of externalized knowledge, e.g., by reading

© The Editor(s) (if applicable) and The Author(s), under exclusive license to 267
Springer-Verlag GmbH, DE, part of Springer Nature 2026
M. Hemmje and S. Wagenpfeil, *Multimedia Information Retrieval,*
https://doi.org/10.1007/978-3-662-73310-3

1.4—Interaction

What is the difference between the HCI and CSCW models?

HCI stands for Human Computer Interaction and describes the interaction of a user with a system. CSCW, on the other hand, stands for Computer-Supported Cooperative Work and describes the interaction between multiple users, where computers are merely considered as tools. The addressee in HCI is thus a machine, while in CSCW it is always people.

1.5—Collections

What fundamental types of collections do you know?

- User-defined collections
- Application-driven collections
- Provided collections
- Dynamic collections

1.6—Terminology

What is the difference between a multimedia object and a multimedia asset?

Compared to a multimedia object, a multimedia asset additionally contains a specific material or intangible value for the user.

1.7—Features

What levels of features do you know?

- Technical features
- General features
- Format-specific features
- Content-based features
- Semantic features
- Application-specific features

A.2 Solutions to Chap. 2

2.1—Text Encoding

Using the US-ASCII character set (see Fig.2.1), determine the hexadecimal encoding of the word "Multimedia".

```
4D 75 6C 74 69 6D 65 64 69 61
```

2.2—Character Sets

What are glyphs?

Glyphs are different manifestations of one and the same character and are often used for representation.

2.3—Character Sets

What is the difference between US-ASCII and Unicode?

Unicode contains a much larger space for characters and can represent almost all internationally used languages and symbols. US-ASCII, on the other hand, is a very small character set and is optimized only for the US region.

2.4—Text Formats

What advantages do XML documents offer?

XML documents are characterized by a clear structure (well-formedness), achieved through the introduction of so-called tags, and by the possibility of validity checking through validation against an XML schema or a document type definition.

2.5—Word Processing

Create a vectorization table for the following text: "There is a tree over there. On the tree sits a bird. The bird is chirping loudly. At the base of the tree sits a cat and watches the bird" (Tab. A.1).

2.6—Index

Create an inverted index at the sentence level for the terms from Exercise 2.5 (Table A.2).

2.7—Index

Determine the term frequency f for the terms from Exercise 2.5. Also consider the relation to the created vectorization table (Tab. A.3).

2.8—Image Features

Explain the term anti-aliasing.
In anti-aliasing, transparency or attenuated color values are used to smooth the edges of certain objects for the human eye, thereby reducing the visibility of individual pixels.

Table A.1 Vectorization table of the example text in the bag-of-words model

Word	Frequency
ein	3
baum	3
vogel	3
sitzt	2
der	2
da	1
drüben	1
steht	1
auf	1
zwitschert	1
laut	1
unten	1
am	1
katze	1
und	1
betrachtet	1

Table A.2 Inverted index of the example text at sentence level

Word	Sentence numbers
am	4
auf	2
baum	1, 2, 4
betrachtet	4
da	1
dem	2
der	3
drüben	1
ein	1, 2
katze	4
laut	3
sieht	–
sitzt	2, 4
steht	1
und	4
unten	4
vogel	2, 3, 4
zwitschert	3

Table A.3 Term frequency (TF) f for selected words in the text—normalized per sentence

Word	Sentence 1	Sentence 2	Sentence 3	Sentence 4
am	0.000	0.000	0.000	0.100
auf	0.000	0.167	0.000	0.000
baum	0.200	0.167	0.000	0.100
betrachtet	0.000	0.000	0.000	0.100
da	0.200	0.000	0.000	0.000
dem	0.000	0.167	0.000	0.000
der	0.000	0.000	0.250	0.000
drüben	0.200	0.000	0.000	0.000
ein	0.200	0.167	0.000	0.000
katze	0.000	0.000	0.000	0.100
laut	0.000	0.000	0.250	0.000
sieht	0.000	0.000	0.000	0.000
sitzt	0.000	0.167	0.000	0.100
steht	0.200	0.000	0.000	0.000
und	0.000	0.000	0.000	0.100
unten	0.000	0.000	0.000	0.100
vogel	0.000	0.167	0.250	0.100
zwitschert	0.000	0.000	0.250	0.000

2.9—Image Features

What is the alpha channel?

A layer of GIF or PNG images that contains transparency information.

2.10—Huffman Coding

Create a Huffman coding for the sentence "Abrakadabra Simsalabim" (Fig. A.1).

a	b	m	i	r	l	s	S		d	k	A
00	010	0110	0111	1000	1001	1010	1011	1100	1101	1110	1111

2.11—Audio Features

Explain the terms sampling and quantization.

Sampling transforms a continuous signal into a series of time-based intervals. Quantization then assigns a discrete value to each element of the interval.

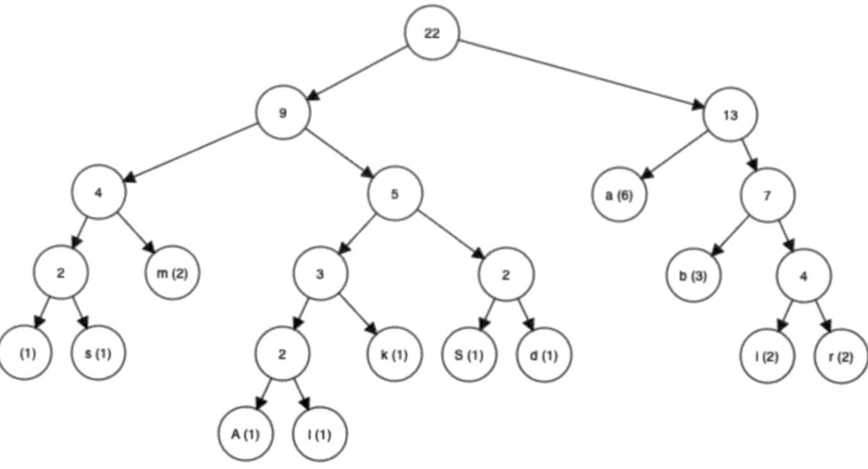

Fig. A.1 Solution to Exercise2.10

2.12—Audio Features

What does the sampling theorem state (briefly explained)?

An analog signal can be digitally reconstructed without loss if it is sampled at a frequency that is at least twice as high as the highest frequency contained in the signal.

2.13 Video Characteristics

Calculate the file size of a video in HD (1280×720), Full HD (1920×1080), and Ultra HD (2840×2160), 60 Hz and a color depth of 24 bit with stereo sound (11 kHz, 8 bit) and a duration of 4:30 min.

- HD: approx. 4 GB
- Full HD: approx. 9 GB
- Ultra HD: approx. 37 GB

2.14—Video Features

What is MPEG-7 used for?

For the description of metadata of videos in a standardized XML format.

2.15—Machine Learning

Which machine learning methods do you know?

- Supervised learning
- Unsupervised learning
- Reinforcement learning

2.16—Graphs

Create a weighted adjacency matrix for Fig. 2.45 under the assumption of the following weights:

$$A \rightarrow B = 5$$
$$A \rightarrow C = 3$$
$$B \rightarrow D = 8$$
$$D \rightarrow C = 3$$
$$C \rightarrow E = 2$$
$$D \rightarrow E = 1$$

$$AM(G_{\text{ex}}) = \begin{pmatrix} 0 & 0 & 0 & 0 & 0 \\ 5 & 0 & 0 & 0 & 0 \\ 3 & 0 & 0 & 3 & 0 \\ 0 & 8 & 0 & 0 & 0 \\ 0 & 0 & 3 & 1 & 0 \end{pmatrix}$$

2.17—Semantics

What is a thesaurus?

A thesaurus is a controlled vocabulary.

2.18—Semantics

What role do functions play in ontologies?

Functions represent relations with constraints.

A.3 Solutions to Chap. 3

3.1—General Tasks

What are the core tasks of MMIR systems?

Ad-hoc retrieval and filtering.

3.2—General Model

What does the RSV value denote?

The Retrieval Status Value indicates the relevance relationship of the results in IR.

3.3—General Model

What are the three conceptual units in the general information model?

- System input
- Processing
- System output

3.4 -Exact-Match Model

Provide the result of the following queries based on the sample document collection (see Table 3.1):

- $t_1 \wedge t_3$
- $t_1 \vee t_3$
- $t_1 \wedge t_2 \vee t_4$
- $t_1 \neg t_2 \wedge t_3$
- $t_1 \vee (t_2 \wedge t_4)$

- $t_1 \wedge t_3 \rightarrow d_1, d_3, d_4$
- $t_1 \vee t_3 \rightarrow d_1, d_2, d_3, d_t$
- $t_1 \wedge t_2 \vee t_4 \rightarrow d_4, d_3$
- $t_1 \neg t_2 \wedge t_3 \rightarrow d_1, d_3$
- $t_1 \vee (t_2 \wedge t_4) \rightarrow d_1, d_3, d_4, d_2$

3.5—Best-Match Models

What distinguishes the vector space model?

The vector space model represents documents and queries as vectors, assigning each term to a dimension. This enables similarity calculations using vector geometry.

3.6—Relevance Feedback

Explain the difference between explicit and implicit relevance feedback.

With explicit relevance feedback, the user directly provides their judgment on the result set. Implicit relevance feedback uses indirect signals (such as click-through rates) to infer the user's relevance judgment.

3.7—Relevance Feedback

Name the five most important strategies for the click-rate method.

- Click > Skip Above
- Last Click > Skip Above
- Click > Earlier Click
- Last Click > Skip Previous
- Click > No-Click Next

3.8—Relevance Feedback

Describe the "Last Click > Skip Previous" strategy.

Here, only the last click is considered to establish a preference. All previous clicks are regarded as irrelevant.

3.9—Relevance Feedback

What other possibilities for relevance feedback do you know?

Reading time, scrolling, interaction

3.10—Query Optimization

What does term co-occurrence mean?

This assumes that terms may have a certain semantic connection and therefore always occur together.

3.11—Persistent Identifiers

Name requirements for persistent identifiers.

- Uniqueness
- Injectivity
- Format support
- Technology independence
- Storage location independence
- Semantic neutrality
- Interoperability
- Language support
- Ease of creation
- Ease of resolution
- Fault tolerance

- Security
- Metadata
- Granularity

3.12—Content Delivery

What role does the delivery node play in content delivery networks?

Delivery nodes are the nodes that deliver content to users.

3.13—Distributed IR

What distinguishes cooperative from non-cooperative environments?

In a non-cooperative environment, a source does not allow access to the data set and associated indexes, whereas a cooperative source would allow this.

3.14—Distributed IR

What is a nugget?

Nuggets are information units that describe the results of distributed queries.

3.15—Distributed IR

What phases does the transfer process of result aggregation have?

- Grouping
- Sorting
- Merging
- Splitting
- Extraction

3.16—Distributed IR

What partitioning options for a collection do you know in distributed IR?

Document-based and term-based

A.4 Solutions to Chap. 4

4.1—Long-term Archiving

Which types of metadata are relevant for long-term archiving?

- Content metadata
- Technical metadata
- Structural metadata
- Administrative metadata
- Rights metadata

4.2—Long-term Archiving

What role does the "Intellectual Entity" play in the PREMIS data model?

It serves to bundle related content that describes a single unit.

4.3—Long-term Archiving

What is a "SIP" in the OAIS reference model?

A Submission Information Package, which is used to represent the information transmitted by producers to the archive.

4.4—Semantics

What characterizes the random walk approach?

In this approach, random paths in a graph are used to calculate semantic similarity.

A.5 Solutions to Chap. 5

4.1—Test Collections

What criteria must be met for test collections according to the Cranfield paradigm?

According to the Cranfield paradigm, a test collection must contain three central components: a defined and representative document collection, a set of clearly formulated test queries that reflect typical information needs, and relevance assessments that specify which documents are considered relevant for which queries. Ideally, these relevance judgments are made by domain experts and serve as an objective basis for evaluation. Only when all three components are present in a consistent and structured manner can information retrieval systems be reliably and comparably evaluated.

4.2—Metrics

How are precision and recall mathematically defined?

$$Recall = \frac{G \cap R}{R}$$

$$Precision = \frac{G \cap R}{G}$$

4.3—Metrics

What is NDCG used for?

It incorporates ranking into the evaluation of results.